Progress in Molecular and Subcellular Biology

Springer

Berlin
Heidelberg
New York
Hong Kong
London
Milan
Paris
Tokyo

Progress in Molecular and Subcellular Biology

Volumes Published in the Series

Alain Beschin · Werner E.G. Müller (Eds.)

Invertebrate Cytokines and the Phylogeny of Immunity

Facts and Paradoxes

With 45 Figures, 3 in Color

Springer

Dr. ALAIN BESCHIN
Department of Cellular and Molecular Interactions
Flemish Interuniversity Institute for Biotechnology
Free University of Brussels (VIB-VUB)
Building E, Floor 8
Pleinlaan 2
1050 Brussels
Belgium

Professor Dr. WERNER E.G. MÜLLER
Institut für Physiologische Chemie
Abt. Angewandte Molecularbiologie
Johannes Gutenberg-Universität
Duesbergweg 6
55099 Mainz
Germany

ISSN 0079-6484
ISBN 3-540-40407-4 Springer-Verlag Berlin Heidelberg New York

Library of Congress Cataloging-in-Publication Data
Invertebrate cytokines and the phylogeny of immunity : facts and paradoxes / Alain Beschin, Werner E.G. Müller.
p. cm. – (Progress in molecular and subcellular biology, ISSN 0079-6484)
Includes bibliographical references.
ISBN 3-540-40407-4 (alk. paper)
1. Invertebrates–Physiology. 2. Cytokines. I. Beschin, Alain. II. Müller, Werner E.G. III. Series.
QH506.P76 no. 0079-6484

Springer-Verlag Berlin Heidelberg New York
a member of BertelsmannSpringer Science+Business Media GmbH
http.//www.springer.de

Cover design: Meta Design, Berlin
Typesetting: SNP Best-set Typesetter Ltd., Hong Kong
39/3150WI – 5 4 3 2 1 0 – Printed on acid-free paper

Preface

Since the pioneering work of E. Metchnikoff at the end of the 19th century, invertebrate immunology has become a proper topic of study. However, it has only been within the last two decades that detailed analyses of invertebrate immune reactions have begun to emerge. These observations in organisms that lack adaptive immunity serve as an insight into the origin of the sophisticated immune networks of vertebrates as well as into the identification of new factors with possible therapeutic uses in human medicine.

Considerations of the functional parallels between innate (non-adaptive) host defenses in invertebrates and vertebrates led to the hypothesis that invertebrates possess soluble cytokine-like mediators that initiate and orchestrate inflammatory responses to invading microorganisms and environmental insults. Until recently, our ability to track the evolutionary lineages of cytokine genes in invertebrates was mainly based on immunocytochemical methods and functional assays. It is only in the last few years that these aspects have been investigated at molecular levels. In the present book, progress lately achieved in research aiming to demonstrate cytokine-like activities and related signaling pathways in protostome and deuterostome invertebrates – as well as their roles in embryogenesis, hematopoiesis and immune response – is critically reviewed taking advantage of the now available complete sequence of the *Caenorhabditis elegans* and *Drosophila* genomes.

Ottaviani et al. (Chap. 1) summarize the functional findings showing that the cellular response of invertebrate species to mammalian cytokines are comparable with those observed in mammals, thus favoring the hypothesis that vertebrate cytokines and their invertebrate functional counterparts have a common phylogenetic origin.

That there is an evolutionary continuity – at the molecular level – of the immune response of invertebrates and vertebrates despite their parallel development for more than 500 million years is further illustrated in several chapters. Bodian and colleagues (Chap. 2) review the molecular conservation of cytokines, growth factors and JAK/STAT signal transduction pathway regulating *Drosophila* and mammalian hematopoiesis. Wajant and Scheurich (Chap. 3) discuss the similarities in the role of TNF receptor-associated factor (TRAF) family members in the activation of conserved intracellular kinase cascades regulating NF-κB- and JNK-dependent defense genes in *Drosophila* and vertebrates. Parker et al. (Chap. 4) provide an overview of ligands belonging to the TGF-β superfamily which includes three subgroups, the TGF-βs, BMPs and activins. They report that prototypical BMPs and activins are present in

Drosophila, being involved in embryonic patterning and larval development. Concomitantly, true TGF-β homologues seem absent, and a role for TGF-β superfamily members in invertebrate immune response has not been demonstrated. These facts therefore suggest that molecules involved in invertebrates embryogenesis may have gained diversity with the emergence of vertebrates, yet being recruited by the immune system. Muller et al. (Chap. 5) focus on intercellular mediators from sponges considered as chemokines based on functional studies revealing that these proteins control the differentiation of stem cells, and mediate immune response and formation of colonial stages. Interestingly, most of the putative sponge chemokines share higher sequence similarities with related proteins from deuterostomes than from protostomes. Some sponge chemokines are even present in deuterostomes while absent in protostomes.

While the previous chapters mainly illustrate a model postulating a common evolutionary origin for vertebrate and invertebrate defense systems, two contributions reveal another – not exclusive – alternative. Beschin et al. (Chap. 6) as well as Raftos and colleagues (Chap. 7) show that unrelated genes can encode protostome and deuterostome invertebrate defense molecules despite functional similarities to vertebrate cytokines, suggesting that new genes may appear and fulfil functions analogous to other known genes through convergent evolution.

It appears on the basis of the current molecular biology findings presented in this book that the level of sequence conservation in the course of evolution may vary from gene to gene. In addition, genes coding for factors involved in intracellular signaling may be more conserved than those encoding extracellular immune mediators. Clearly, the coming years will be a major time for gene discovery in most invertebrates, and the track of evolutionary relationships between innate defense genes in vertebrates and invertebrates will continue to benefit from the availability of the *Drosophila* genome. However, a better understanding of evolutionary origins of innate defense genes will only be possible by extending substantially the collection of organisms in which the genes have been characterized. In this regard, the observation that immune defense molecules of sponges, the evolutionary lowest multicellular animals, are more closely related to vertebrates (deuterostomes) than to invertebrates (protostomes) deserves consideration. Accordingly, the ongoing sequencing of sponge genomes (*Suberites domuncula*, *Geodia cydonium*, ...), the now completed genome of the sea squirt *Ciona intestinalis*, a chordate which represents an important branch point in the prevertebrate evolution, and the further completion of the genome of the sea urchin *Strongylocentrotus purpuratus* occupying a unique position in animal evolution as the only non-chordate invertebrate deuterostome, will provide invaluable out-groups to study the evolution of cytokine-related genes in vertebrates.

I would like to take this opportunity to thank all the authors who made this book possible.

ALAIN BESCHIN

Contents

Analogies Between *Drosophila* and Mammalian TRAF Pathways
H. Wajant and P. Scheurich

Regulation of BMP and Activin Signaling in *Drosophila*
L. Parker, D.G. Stathakis, and K. Arora

The Chemokine Networks in Sponges: Potential Roles in Morphogenesis, Immunity and Stem Cell Formation
W.E.G. Müller, M. Wiens, I.M. Müller, and H.C. Schröder

Invertebrate Humoral Factors: Cytokines as Mediators of Cell Survival

E. Ottaviani, D. Malagoli, and A. Franchini

Abstract. The presence and the different functional aspects of cytokine-related molecules in invertebrates are described. Cytokine-like factors affect immune functions, such as cell motility, chemotaxis, phagocytosis and cytotoxicity. In particular, cell migration shows a species-specific effect for IL-1α and TNF-α and a dose-correlated effect for IL-8, PDGF-AB and TGF-β1. Apart from some exceptions, the phagocytic effect increases significantly at all the concentrations tested and with all the species used. PDGF-AB, TGF-β1 and IL-8 provoke conformational changes in mollusk immunocytes, involving the signaling transduction pathways of phosphatidylinositol and cAMP. PDGF-AB and TGF-β1 partially inhibit the induced programmed cell death in an insect cell line, and the survival effect is mediated by the activation of phosphatidylinositol 3-kinase, PKA and PKC. The exogenous administration of these growth factors in an invertebrate wound repair model showed that they are able to control the wound environment and promote the repair process by accelerating the coordinated activities involved. Moreover, IL-1α, IL-2 and TNF-α are able to induce nitric oxide synthase. PDGF-AB and TGF-β1 provoke an increase in neutral endopeptidase-24.11 (NEP)-like activity in membrane preparations from mollusk immunocytes, while NEP deactivates the PDGF-AB- and TGF-β1-induced cell shape changes. Cytokines are also involved in invertebrate stress response in a manner extremely similar to that in vertebrates. Several studies suggest the existence on the mollusk immunocyte membrane of an ancestral receptor capable of binding both IL-2 and CRH. Furthermore, the competition found between CRH and a large number of cytokines supports the idea that invertebrate cytokine receptors show a certain degree of promiscuity. The multiple functions of cytokines detected in invertebrates underline another characteristic of mammalian cytokines, i.e. their great pleiotropicity. Altogether, the studies on the function of the invertebrate humoral factors show a close overlapping with those found in vertebrates, and the hypothesized missing correlation between invertebrate and vertebrate cytokine genes that is emerging from the limited molecular biology data present in literature might represent a very peculiar strategy followed by Nature in the evolution of cytokines.

1 Introduction

Cytokines were first discovered by pioneer researchers in the immunology field. However, it has recently become apparent that these molecules are also primary mediators of a variety of physiological functions in non-immune organs and systems. These molecules are a large number of different types of glycoproteins which often share common functions. They are produced by

Department of Animal Biology, University of Modena and Reggio Emilia, Via Campi 213/D, 41100 Modena, Italy

Progress in Molecular and Subcellular Biology
A. Beschin (Ed.) Invertebrate Cytokines

both activated immune cells and cells belonging to other systems. Cells of the neuroendocrine system are also able to produce cytokines that play an important role as signal molecules (Nisticò 1993; Blalock 1994; Schöbitz et al. 1994). The list of identified and cloned cytokines is very long and still growing. They have been given different names according to their origin or function. In mammals, the main examples are: interleukins (IL), tumor necrosis factor (TNF), interferons (IFN) and transforming growth factor (TGF)-β. Cytokine-like molecules have also been found in different tissues of various invertebrate species (Table 1, Fig. 1).

It should be underlined that the immune system shows biological characteristics that are also representative of other systems, i.e. the nervous and endocrine systems. These systems share low molecular weight signals that mediate distant interactions (cytokines, neurotransmitters, hormones), high molecular weight signals that mediate close cell-cell interactions (receptors), and finally, the driving forces responsible for cellular movement, proliferation and programmed cell death. In this light, the distinction between cytokines, neurotransmitters and hormones becomes merely semantic, as the same substance may belong to one or more systems, suggesting that Nature has pursued a similar strategy in the development of the immune and neuroendocrine systems. Following a phylogenetic approach, we have surmised a common origin of the immune and neuroendocrine systems (Ottaviani and Franceschi 1997). From the beginning of evolution, a common pool of molecules, such as cytokines, proopiomelanocortin (POMC)-products and neuropeptides, has been used to provide an integrated network of adaptive mechanisms that are critical for survival.

Among cytokines, growth factors regulate several processes, such as cell proliferation, differentiation, matrix production, wound repair and survival that underwrite animal embryo development and tissue homeostasis (Mustoe et al. 1987; Cross and Dexter 1991; Kresse and Schonherr 2001).

Recently, studies on the IPLB-LdFB cell line from the insect *Lymantria dispar* have shown that platelet-derived growth factor (PDGF)-AB and TGF-β1 partially inhibit the programmed cell death induced by a classical apoptosis inducer of mammalian cells, i.e. the reducing sugar 2-deoxy-D-ribose which acts by oxidative stress (Table 2). Indeed, the sugar effect was reversed by *N*-acetyl-L-cysteine, which is considered a glutathione precursor and a scavenger of free radicals (Staal et al. 1992; Buttke and Sandstrom 1994; Mayer and Noble 1994; Ottaviani et al. 2000a, 2001). The survival effect of PDGF-AB and TGF-β1 is mediated by the activation of phosphatidylinositol 3-kinase, protein kinase A (PKA) and protein kinase C (PKC) (Ottaviani et al. 2001). The inhibiting effect of growth factors as well as their modality of action are widely reported in mammalian cells (Collins et al. 1994).

Furthermore, the exogenous administration of PDGF-AB and TGF-β1 in an invertebrate wound repair model showed that they control the wound environment and promote the repair process by accelerating the coordinated activities involved (Franchini and Ottaviani 2000). In this study, we demonstrated

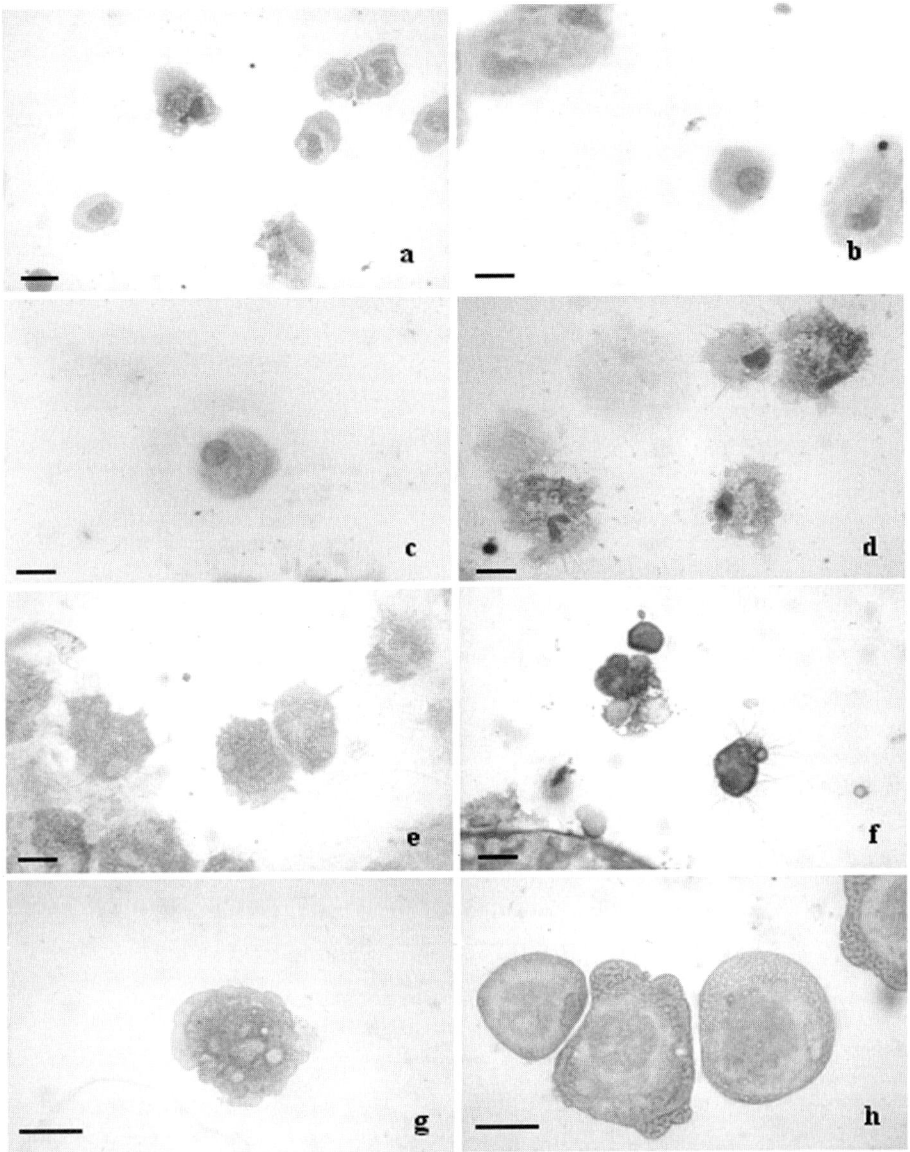

Fig. 1. Immunoperoxidase labeling of immunocytes of different species with anti-cytokine antibodies. Cell nuclei were counterstained with hematoxylin. **a** *Planorbarius corneus*: anti-PDGF-AB polyclonal antibody (pAb) ; **b** *Viviparus ater*: anti-IL-1α pAb and **c** anti-IL-6 monoclonal antibody (mAb) ; **d** *Mytilus galloprovincialis*: anti-TGF-β1 pAb and **e** anti-IL-8 mAb; **f** *Calliphora vomitoria*: anti-TNF-α pAb; **g** *Eisenia foetida*: anti-TGF-β1 pAb; **h** *Lymantria dispar*: anti-TGF-β1 pAb. *Bar* = 10 μm

Table 1. Cytokine-like molecules found in various invertebrate taxa

Taxa	Cytokine-like molecules	References
Mollusca		
Mytilus edulis	IL-1α,β IL-6 TNF-α	Hughes et al. (1990, 1991, 1992); Stefano et al. (1991); Paeman et al. (1992)
Planorbarius corneus	IL-1α,β IL-2 IL-6 TNF-α PDGF-AB TGF-β1	Ottaviani et al. (1993b); Franchini et al. (1996a)
Viviparus ater	IL-1α,β IL-2 IL-6 TNF-α PDGF-AB TGF-β1	Ottaviani et al. (1993b); Franchini et al. (1996a)
Biomphalaria glabrata	IL-1 TNF-α	Granath et al. (1994); Owe-Missi-Oukem-Boyer et al. (1994)
Viviparus contectus	PDGF-AB TGF-β1	Franchini et al. (1996a)
Lymnaea stagnalis	PDGF-AB TGF-β1 EGF neurotrophic factor	Fainzilber et al. (1996); Franchini et al. (1996a); Hermann et al. (2000)
Mytilus galloprovincialis	IL-8 PDGF-AB TGF-β1	Franchini et al. (1996a); Ottaviani et al. (2000b)
Crassostrea gigas	TGF-β	Lelong et al. (2000)
Insecta		
Bombyx mori	Growth promoting factor	Aizawa and Sato (1963); Vaughn and Louloudes (1978)
Samia cynthia	Growth promoting factor hemokinin	Williams and Kambysellis (1969); Cherbas (1973)
Antheraea polyphemus	Hemokinin	Cherbas (1973)
Hyalophora cecropea	Hemokinin	Cherbas (1973)
Drosophila melanogaster	EGF TGF-α,β neurotrophic factor imaginal disc GFs	Wharton et al. (1985); Kelley et al. (1987); Padgett et al. (1987); Bryant (1988); Kopczynski et al. (1988); Hayashi et al. (1992); Neuman-Silberberg and Schupback (1993); Kutty et al. (1998); Kawamura et al. (1999)
Manduca sexta	Hemolymph trophic factor	Wielgus et al. (1990)
Calliphora vomitoria	TNF-α PDGF-AB TGF-β1	Franchini et al. (1996a,b)
Galleria mellonella	IL-1α TNF-α	Wittwer et al. (1999)
Estigmene acraea	IL-1α TNF-α	Wittwer et al. (1999)
Lymantria dispar	PDGF-AB TGF-β1	Ottaviani et al. (2000a)
Nematoda		
Caenorhabditis elegans	EGF FGF TGF-β IL-1α TNF-α PDGF-AB	Hill and Sternberg (1992); Ren et al. (1996); Burdine et al. (1997); Roubin et al. (1999); Suzuki et al. (1999); Franchini et al. (2003)
Brugia malayi	TGF-β	Gomez-Escobar et al. (1998)
Brugia pahandi	TGF-β	Gomez-Escobar et al. (1998)
Annelida		
Nereis diversicolor	IL-1α	Paeman et al. (1992)
Eisenia foetida	IL-1α TNF-α PDGF-AB TGF-β1	Cooper et al. (1995); Franchini et al. (1996a)

Table 1. *Continued*

Taxa	Cytokine-like molecules	References
Echinodermata		
Asterias forbesi	IL-1 IL-6	Beck and Habicht (1986, 1994)
Pisaster ochraceus	IL-1α	Burke and Watkins (1991)
Anthocidaris crassispina	EGF	Yamasu et al. (1995)
Lytechinus pictus	PDGF-B TGF-α	Govindarajan et al. (1995)
Strongylocentrotus purpuratus	TGF-β	Stenzel et al. (1994); Ponce et al. (1999)
Tunicata		
Botryllus schlosserii	IL-1 IL-1α TNF-α	Beck et al. (1989); Ballarin et al. (2001)
Styela placata	IL-1	Beck et al. (1989)
Molgula occidentalis	IL-1	Beck et al. (1989)
Ciona intestinalis	IL-1 EGF	Beck et al. (1989); Nakayama et al. (2001)
Amaroucium pellucidum	IL-1	Beck et al. (1989)
Styela clava	IL-1	Beck et al. (1989)
Halocynthia roretzi	TGF-β	Miya et al. (1996); Darras and Nishida (2001)
Boltenia villosa	EGF	Davidson and Swalla (2001)
Molgula pacifica	FGF	Bates (1998)

Table 2. Percentage of apoptotic *Lymantria dispar* IPLB-LdFB cell line following the addition to the cell medium of 20 ng/ml PDGF-AB, 5 pg/ml TGF-β1, 50 mM 2-deoxy-D-ribose (dRib), 50 mM dRib plus PDGF-AB or 50 mM dRib plus 5 pg/ml TGF-β1. (Ottaviani et al. 2000a, reprinted with permission of the publisher)

Substances	Mean ± SD
Control	6.03 ± 0.98
PDGF-AB	6.06 ± 0.23
TGF-β1	7.40 ± 0.17
dRib	31.57 ± 2.82*
dRib+PDGF-AB	19.30 ± 2.59**
dRib+TGF-β1	20.23 ± 0.23**

Statistical analysis was performed by ANOVA. *$P < 0.05$ vs. control; **$P < 0.05$ vs. dRib.

that PDGF-AB and TGF-β1 enhance the speed of the healing process of the mollusk *Limax maximus*, and are involved in all the successive phases. First, the infiltration phase is activated and cell migration into the wound area stimulated. In particular, an increased number of immunocytes is seen to stratify at the margins of the injury site, while a larger, closing area of clumped cells is observed, together with a reduction in the damaged tissue by the activation of immunocyte phagocytic functions. The subsequent formation and maturation of granulation tissue is also stimulated by an earlier development of hematic lacunae and greater deposition of extracellular matrix component by fibroblast function and immunocyte fibroblast-like activity. The re-epithelialisation of the wound area is accelerated in comparison to the control incisions. The growth factors also stimulate the transition of marginal wound epithelial cells to a migratory phenotype in order to cover the repairing area and restore the epithelium function more quickly. No significant differences in the healing properties of the two growth factors were observed. In mammals, growth factors, and in particular PDGF and TGF-β, released locally by immunocytes or by adjacent somatic cells have been demonstrated to be key players in the healing process, as they are potent stimulants of granulation tissue formation (Clark 1996). Furthermore, exogenous addition of these factors seems to accelerate the repair process (Grotendorst et al. 1985; Roberts et al. 1986; Pierce et al. 1989, 1992; Quaglino et al. 1991; Khouri et al. 1993).

Lastly, we are performing studies on the role of cytokines in aging. In young adults (fertile stage: 3 days after L1 stage) of the nematode *Caenorhabditis elegans*, IL-1α and PDGF-AB immunoreactive molecules have been found in neurons, and TNF-α-like molecules in the secretory cells of the pharyngeal terminal bulb (Franchini et al. 2003). A different pattern was observed in old adult (unfertile stage: 12 days after L1 stage) worms, where a general decrease in the number of cells positive to the antibodies tested was found, i.e. disappearance of PDGF-AB positivity and a less intensity for TNF-α. Studies in mammals have shown that cytokines may contribute to age-associated changes. With regard to the role of cytokines in aging, Franceschi et al. (1995) refer to "the 'yo-yo' cytokine network".

2 Immune Response

Phagocytosis and cytotoxicity are basic defense mechanisms in invertebrates. As these functions are so important for cell survival, they are very well preserved through evolution and integrated into more sophisticated types of response which appeared later in evolution, such as acquired immunity by clonotypically distributed molecules, i.e. antibodies. The innate invertebrate immune response involves both cellular and humoral components. Regarding the latter, several molecules, including cytokine-like factors, have been described in invertebrates (for review see Ratcliffe et al. 1985), and ever more data on the role of these molecules in immune functions is emerging. Our and other laboratories have reported that they affect cell motility (cell shape changes) and chemotaxis (non-random migration). The latter allows cells to

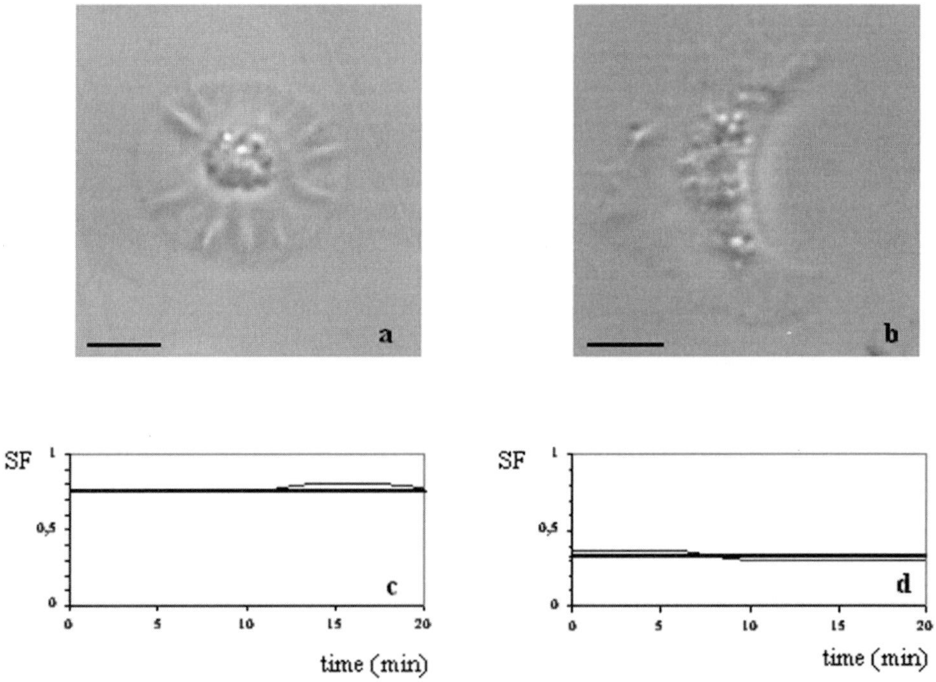

Fig. 2. Effect of IL-8 on *Mytilus galloprovincialis* immunocyte cell shape by image analysis. Control: **a** incubation with marine solution; **b** immunocyte activated in the presence of 100 ng/ml IL-8; the immunocyte cellular shape expressed as shape factor (*SF*) at different times decreased **d** in the presence of IL-8 compared to **c** control. *Bar* = 5 μm

move towards a chemoattractant that has to be recognized. When the chemoattractant is a particulate material such as bacteria, chemotaxis may be followed by engulfment and phagocytosis. PDGF-AB, TGF-β1 and IL-8 are able to provoke changes in immunocyte shape, expressed as modifications in cell shape factor, a parameter which evaluates the morphological changes in cellular shape from an inactive (round form) to an active state (amoeboid form) by means of computer-assisted microscopic image analysis (Fig. 2; Kletsas et al. 1998; Ottaviani et al. 2000b). Furthermore, the signaling transduction pathways involved in conformational changes have been studied using different inhibitors. The use of verapamil, an L-type Ca^{2+}-channel inhibitor, and ω-conotoxin, an N-type Ca^{2+}-channel inhibitor, demonstrated that PDGF-AB induction of the cell migration of invertebrate immunocytes is extracellular Ca^{2+}-independent, while that of TGF-β1 is Ca^{2+}-dependent. Moreover, the plasma membrane N-type Ca^{2+}-channels seem to be important for the action of TGF-β1 (Ottaviani et al. 1997c). The phosphatidylinositol and cAMP pathways are involved in the induction of morphological changes by PDGF-AB and TGF-β1 (Kletsas et al. 1998). At the concentration of 0.6 mM neomycin sulfate,

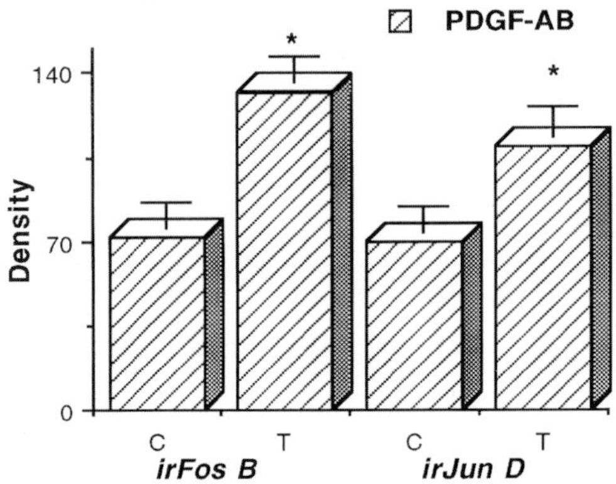

Fig. 3. Densitometric quantification of *Mytilus galloprovincialis* immunocytes immunostained with anti-Fos B and anti-Jun D antibodies after the addition of 20 ng/ml PDGF-AB. *C* Control samples; *T* samples incubated with PDGF-AB; *ir* immunoreactive; $^{*}P < 0.05$ (Student's *t*-test). (Ottaviani et al. 1999, reprinted with permission of the publisher)

a partial inhibition of PDGF-AB and TGF-β1 effects was observed, while at a concentration of 1.5 mM, the inhibition was completed (Kletsas et al. 1998). It should be remembered that at the low concentration used, neomycin blocks the release of phosphatidylinositol 4,5-bis-phosphate (IP_3), while at the higher concentration, the inhibitor blocks not only IP_3, but also the release of diacylglycerol (Carney et al. 1985). It is interesting to note the surprising partial effect of 0.6 mM neomycin on TGF-β1 that is extracellular Ca^{2+}-dependent (Ottaviani et al. 1997c; Kletsas et al. 1998). Taking into account that the PKC activation process implies its Ca^{2+}-mediated translocation from the cytosol to the membrane (Berridge 1987), we can surmise the presence of PKC Ca^{2+}-independent isoforms in our invertebrate model (Hug and Sarre 1993). A PKC activation in the absence of an increase in free cytosolic Ca^{2+} has been reported by di Virgilio et al. (1984) in human neutrophils and Harris et al. (1985) in rat pituitary cells. Furthermore, staurosporine, a potent PKC and PKA antagonist, completely inhibits both PDGF-AB- and TGF-β1-induced cell shape changes (Ottaviani et al. 1999). However, a different action was observed with specific protein kinase inhibitors. Calphostin C (CC), a selective PKC inhibitor (Tamaoki 1991; Nishikawa et al. 1996), completely blocked the action of TGF-β1 but provokes a partial action of PDGF-AB, while H-89, a specific inhibitor of PKA (Chjiwa et al. 1990; Wieprecht et al. 1994), was able to prevent the PDGF-AB- and TGF-β1-induced morphological changes in immunocytes. Following the growth factor-mediated signaling downstream, we have found that the Fos B-, Jun D- and Smad4-like molecules present in the nucleus of the immunocytes were up-regulated after PDGF-AB or TGF-β1 stimulation (Figs. 3, 4; Ottaviani et al. 1999). On the whole, PDGF-AB- and TGF-β1-induced cell shape changes through both PKA and PKC pathways, however the latter is prevalent

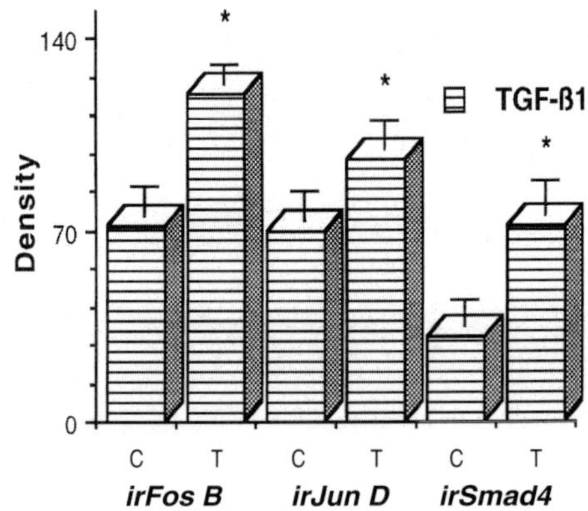

Fig. 4. Densitometric quantification of *Mytilus galloprovincialis* immunocytes immunostained with anti-Fos B, anti-Jun D and anti-Smad4 antibodies after the addition of 5 pg/ml TGF-β1. *C* Control samples; *T* samples incubated with TGF-β1; *ir* immunoreactive; $*P < 0.05$ (Student's *t*-test). (Ottaviani et al. 1999, reprinted with permission of the publisher)

(Kletsas et al. 1998; Ottaviani et al. 1999). The opposite behavior was found for IL-8 (Ottaviani et al. 2000b), where the cAMP pathway prevails over the phosphoinositide one. The IL-8 action was partially inhibited by H-89 and CC, while 2′5′-dideoxyadenosine, which inhibits adenylate cyclase activity (Holgate et al. 1980), completely blocked the chemokine effect. Immunofluorescence studies of the effect of IL-8 on immunocyte shape modifications revealed a time-dependence in the reorganization of the pattern of the actin microfilaments. The incubation of cells in a humidified chamber for 1 min followed by staining with FITC-phalloidin showed positivity localized at the cell periphery and, in particular, distributed in thin, radially expanded pseudopods. After 3–5 min, the immunocytes began to withdraw the pseudopodia and to change shape, assuming a round form due to cell adhesion, and the phalloidin positivity was only concentrated at the periphery under the plasma membrane. The immunocytes incubated with IL-8 showed conformational changes, as cells modified from a round to a polarized form. The maximum modification was detected after 5 min treatment and was maintained for 10 min. The cell shape changes involved a redistribution of actin. Phalloidin positivity was not uniformly distributed at the periphery, but concentrated in few areas where the cells contacted the substrate (Ottaviani et al. 2000b).

All the data indicate that cytokines such as PDGF-AB, TGF-β1 and IL-8 induce cell shape changes in molluscan immunocytes through multiple signal transduction pathways that are well conserved during evolution from invertebrates to vertebrates.

Experiments performed using microchemotaxis chambers showed that IL-1α, TNF-α, IL-8, PDGF-AB and TGF-β1 affect molluscan immunocyte migration (Fig. 5). However, the chemotactic action of these molecules is unclear, because discordant effects were observed. In particular, IL-1α and TNF-α effects were species-specific (Ottaviani et al. 1995b), while IL-8, PDGF-AB and

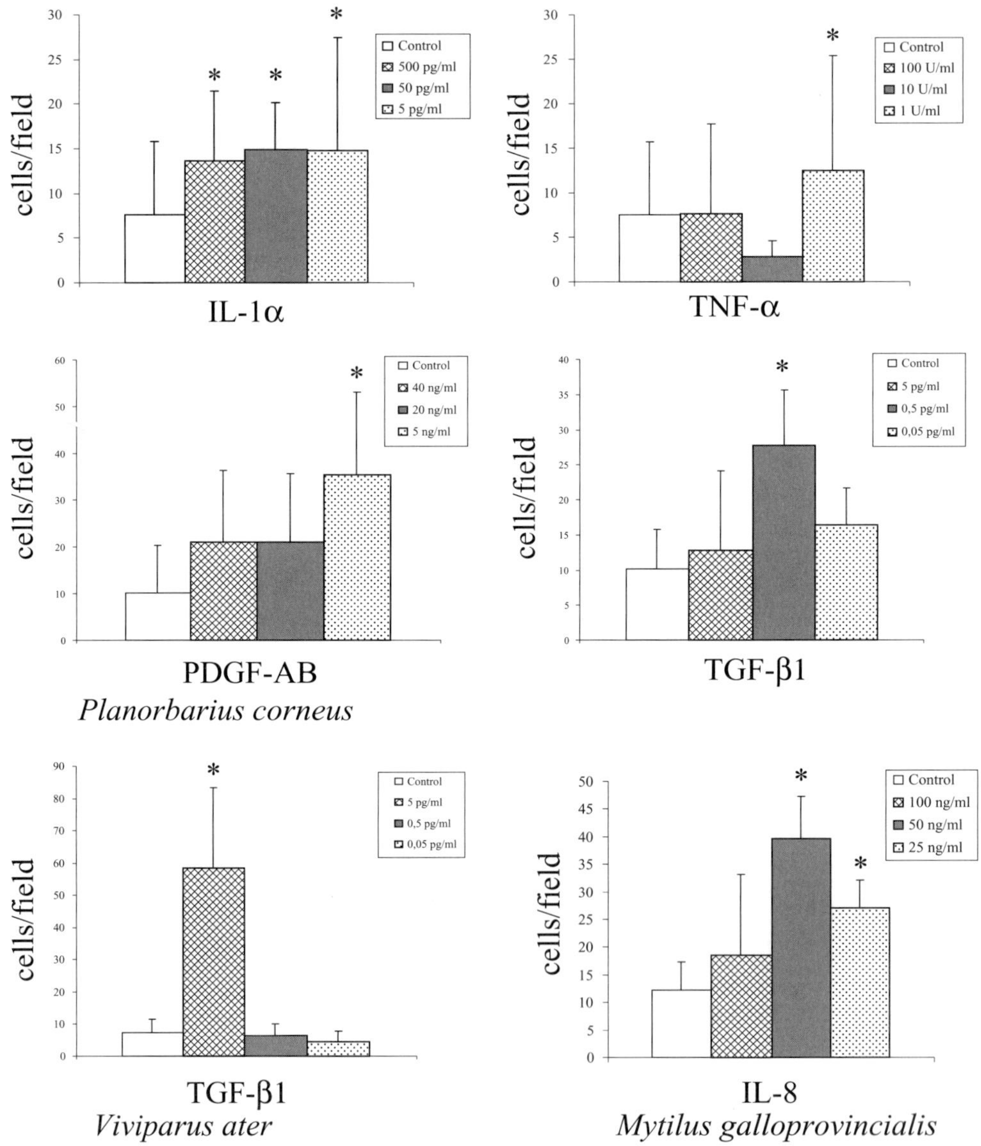

Fig. 5. Influence of various cytokines on the cell migration of immunocytes from molluscan species. *$P < 0.05$ (Student's *t*-test)

TGF-β1 effects were dose-correlated (Ottaviani et al. 1997b, 2000b). The effect on phagocytosis was more uniform (Fig. 6). Indeed, apart from the behavior of PDGF-AB and TGF-β1, the other cytokines showed a significant increase in phagocytic activity at all the concentrations tested and with all the species used (Ottaviani et al. 1995b, 1997b, 2000b). Moreover, the general assumption in vertebrates that chemotaxis and phagocytosis are correlated seems not to hold in invertebrate models. For instance, IL-2 increases the bacterial phagocytosis, but is ineffective on cell migration (Ottaviani et al. 1995b).

In two molluscan species, it has been demonstrated that IL-1α, IL-2 and TNF-α are able to induce nitric oxide synthase (NOS) (Fig. 7; Ottaviani et al. 1995b). The maximum induction was observed 40 min after cytokine injection, and the effect was annulled after 90 min. Furthermore, it has also been found that bacteria-activated immunocytes provoke the induction of this enzyme (Franchini et al. 1995). From a functional point of view, the elimination of bacteria occurs by phagocytosis and the nitric oxide (NO) system. NO induces the bacterial clumping around immunocytes and the subsequent killing, suggesting that the NO system is not alternative to phagocytosis, but that both are involved in the immunocyte response (Ottaviani et al. 1993c; Franchini et al. 1995).

Neutral endopeptidase-24.11 (NEP) is a cell surface zinc metallopeptidase of 749 amino acids known to cleave preferentially small active peptides. A degradative action on cytokines has also been reported (Pierart et al. 1988; Kimura et al. 1991; Casey et al. 1993; Connelly et al. 1993; Kondepudi and Johnson 1993; Delikat et al. 1994; Tharaux et al. 1997; van der Velden et al. 1998). We have found that PDGF-AB and TGF-β1 provoke an increase in NEP-like activity in membrane preparations from immunocytes of the mollusk *Mytilus galloprovincialis* (Table 3), while NEP deactivates the PDGF-AB- and TGF-β1-induced shape changes in immunocytes (Table 4), suggesting that the interaction between cytokines and NEP could modulate immunocyte homeostasis (Caselgrandi et al. 2000).

Table 3. Neutral endopeptidase-24.11 (NEP) membrane bound form-like activity in *Mytilus galloprovincialis* immunocytes in the presence of PDGF-AB and TGF-β1. (Caselgrandi et al. 2000, reprinted with permission of the publisher)

Substances	Mean ± SD
Control	7.83 ± 2.08
PDGF-AB	19.85 ± 1.20*
PDGF-AB+Pho	7.57 ± 1.74
TGF-β1	43.00 ± 5.75*
TGF-β1+Pho	6.28 ± 0.13

Data are express as pmol/10^6 cells. Pho, Phosphoramidon. Statistical analysis was performed by ANOVA. *$P < 0.05$ vs. control.

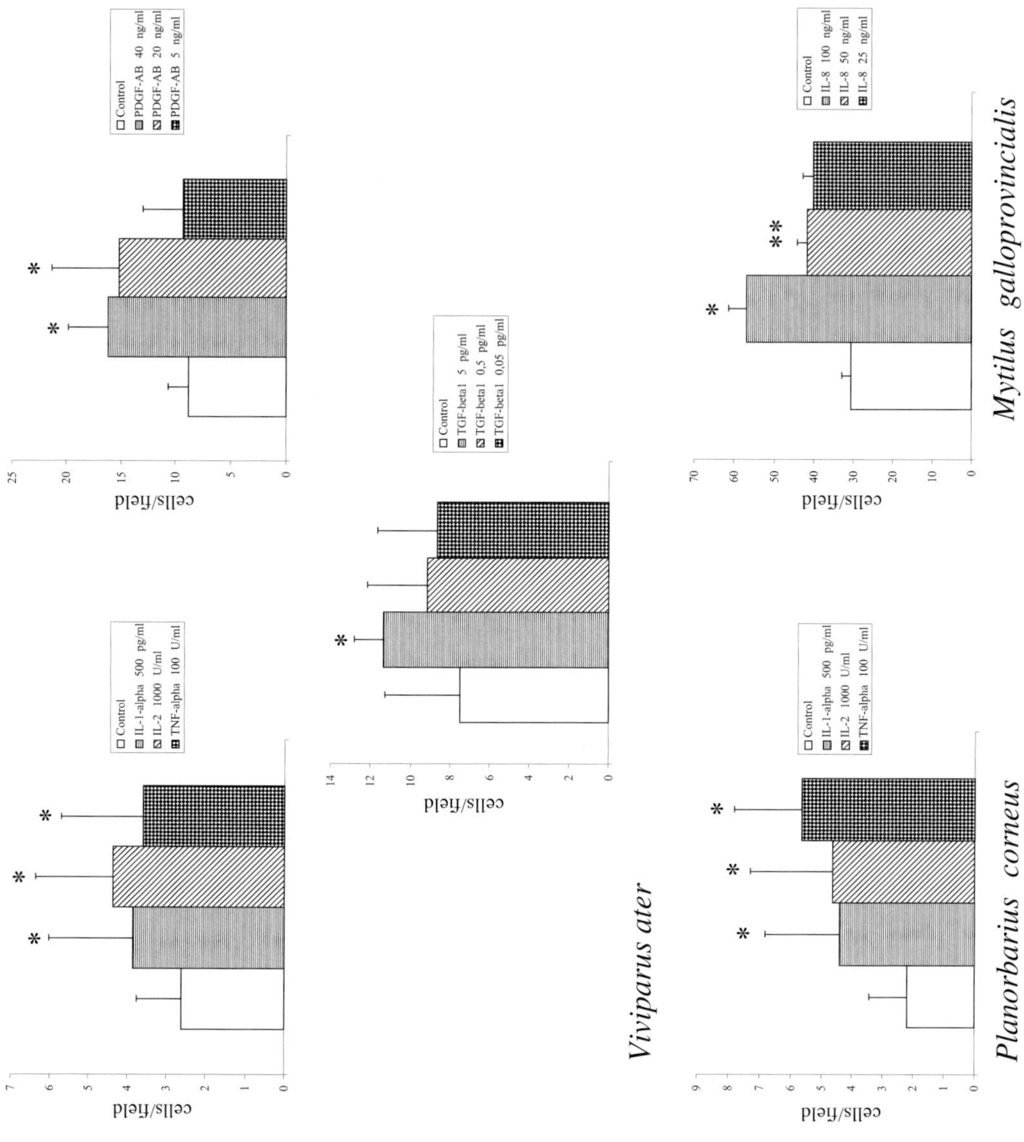
cells/field
Control
IL-1-alpha 500 pg/ml
IL-2 1000 U/ml
TNF-alpha 100 U/ml
PDGF-AB 40 ng/ml
PDGF-AB 20 ng/ml
PDGF-AB 5 ng/ml
TGF-beta1 5 pg/ml
TGF-beta1 0,5 pg/ml
TGF-beta1 0,05 pg/ml
IL-8 100 ng/ml
IL-8 50 ng/ml
IL-8 25 ng/ml
Viviparus ater
Planorbarius corneus
Mytilus galloprovincialis

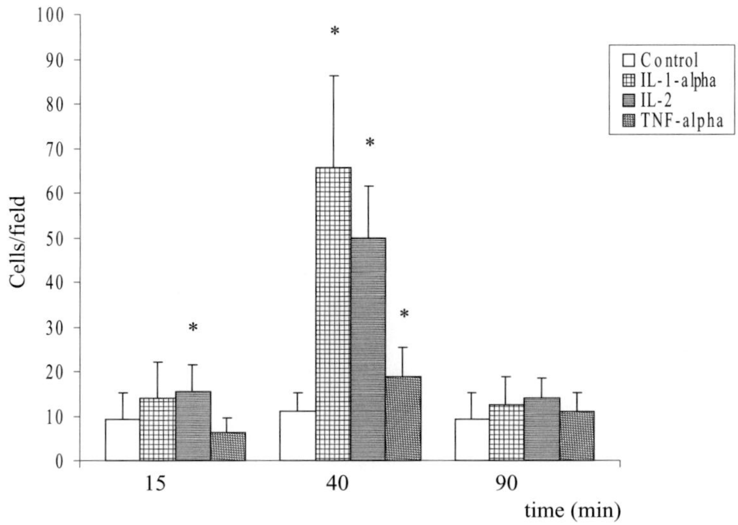

Fig. 7. Induction of nitric oxide synthase (NOS) by cytokines in *Viviparus ater* immunocytes. $*P < 0.05$ (Student's t-test). (Modified from Ottaviani et al. 1995b)

Table 4. Effects of PDGF-AB and TGF-β1 on *Mytilus galloprovincialis* immunocyte shape changes, expressed as shape factor (SF), following incubation with the indicated mixtures. (Caselgrandi et al. 2000, reprinted with permission of the publisher)

Substances	Mean ± SD
Control	0.65 ± 0.01
PDGF-AB	0.27 ± 0.02*
PDGF-AB+NEP[1]	0.67 ± 0.02
PDGF-AB+NEP[2]	0.66 ± 0.03
PDGF-AB+NEP[1]+Pho	0.29 ± 0.03
TGF-β1	0.30 ± 0.03*
TGF-β1+NEP[1]	0.64 ± 0.02
TGF-β1+NEP[2]	0.65 ± 0.02
TGF-β1+NEP[1]+Pho	0.29 ± 0.03

PDGF-AB (20 ng/ml); NEP[1] (2.5 ng/ml); NEP[2] (0.25 ng/ml); TGF-β1 (5 pg/ml); phosphoramidon (Pho) (10 μM). Statistical analysis was performed by ANOVA. $*P < 0.05$ vs control.

◀

Fig. 6. Influence of various cytokines on the phagocytic activity of immunocytes from molluscan species. $*P < 0.05$ (Student's t-test)

Cytotoxicity is another very important defense mechanism which preserves body integrity by killing intruding cells. It is used both under physiological and pathological conditions. This immune function is well preserved through evolution. As reviewed by Ratcliffe et al. (1985), it is present in a wide range of invertebrate species from sponges and coelenterates to sipunculidis and annelids, mollusks, arthropods, echinoderms and protochordates. Different cytotoxic modalities are described in these forms of life. We have demonstrated a relationship between cytokines and mollusk cytotoxic activity (Franceschi et al. 1991). One of the two cell types present in the hemolymph of the mollusk *Planorbarius corneus*, i.e. round hemocyte (RH), has characteristics reminiscent of lymphocyte lineage and reacts with several mouse anti-human monoclonal antibodies, including those directed against epitopes typical of mammalian NK cells. The RH is able to exert a cytotoxic activity on the K562 cell line in a short-term (4 h) ^{51}Cr release test used for evaluating human NK cell activity, while the activity is substantially reduced after 18 h incubation at 24 °C. The presence of human IL-2 during the incubation period preserved the NK-like activity of RH, while none of the other cytokines tested, including IFN-γ, TNF-α, β and IL-4 was able to produce a similar effect (Table 5).

Several groups have identified cytokine-like activities in invertebrates comparable to those in vertebrates. Stefano's group (Hughes et al. 1990, 1991; Stefano et al. 1991) reported that the immunocytes of *Mytilus edulis* respond to IL-1 and TNF-α in a similar way to human granulocytes. As in mammals, IL-1 can induce the formation of TNF-α-like molecules in immunocytes. IL-1 and TNF-α provoke morphological cellular changes in immunocytes, and this action is inhibited by the corresponding antibodies. Lipopolysaccharide (LPS) activates about 92–96% of the immunocytes, and the simultaneous addition of anti-IL-1 and anti-TNF-α antibodies reduces the stimulation of the cells to 50%. From these findings, the authors hypothesize that *M. edulis* immunocytes may produce or release IL-1- and TNF-α-like molecules in response to LPS. Furthermore, the nervous and immune systems in *M. edulis* have been seen to communicate through an opioid–IL-1 relationship. Granath and colleagues (Granath et al. 1994; Connors et al. 1995) found that human IL-1 stimulated the phagocytosis and the production of superoxide in the immunocytes

Table 5. Effect of different cytokines on the NK-like activity immunocytes from the mollusk *Planorbarius corneus*. (Modified from Franceschi et al. 1991)

Cell type[a]	Temperature	Control	TNF-α	TNF-β	IFN-γ	IL-2	IL-4
	(°C)		(5600 U/ml)	(3500 U/ml)	(5000 U/ml)	(20 U/ml)	(2000 U/ml)
RH	24	3.2	3.7	5.8	4.0	20.1	3.2
SH	24	0.5	0.1	2.4	0.1	0.0	0.7

Data are referring to the percentages of specific lysis of the human cell line K562 in a 4 h ^{51}Cr release assay. The E/T cell ratio was 25/1. The temperature refers to the incubation with the cytokines.
[a]RH, Round hemocyte; SH, spreading hemocyte.

of the mollusk *Biomphalaria glabrata*. Furthermore, they suggest that IL-1-like activity in *B. glabrata* could be involved in the resistance to *Schistosoma mansoni* infection. A similar role has also been predicted for TNF-α-like molecules (Ouwe-Missi-Oukem-Boyer et al. 1994). With regard to protochordates, Prendergast's group (Prendergast and Suzuki 1970; Prendergast et al. 1983) first demonstrated the presence of a humoral factor called the "seastar factor" that shows features characteristic of vertebrate cytokines in the echinoderm *Asterias forbesi*. Subsequently, Beck and colleagues (Beck et al. 1989, 1993; Raftos et al. 1991; Beck 1998) isolated and characterized IL-1-, IL-6- and TNF-α-like molecules in *A. forbesi* and the tunicate *Styela clava*. These invertebrate cytokines share many biological activities with their vertebrate counterparts.

However, together with the presence of similar cytokine activity patterns both in vertebrates and invertebrates, peculiar functions of these molecules have been described in invertebrate species. When recruited into the capsule formation, the immunocytes of the fly *Calliphora vomitoria* express TNF-α-like molecules, suggesting an induction of these molecules, which by virtue of their chemotactic properties must act as a chemoattractant between the capsule and the immunocytes (Franchini et al. 1996b). A defense complex consisting of an IL-1-like molecule and the phenoloxidase, an enzyme responsible for the melanization reaction, has been reported in the hemolymph of the insect *Manduca sexta* (Beck et al. 1996). The stimulation of immunocytes of the ascidian *Botryllus schlosseri* with mannan or phorbol 12-mono-myristate induced a marked and significant increase in IL-1α- and TNF-α-like molecules only in morula cells, suggesting that these cells are the main source of these factors in *B. schlosseri* hemolymph (Ballarin et al. 2001).

3
Stress Response

The role of POMC-derived peptides and corticotrophin-releasing hormone (CRH) as central mediators of the stress response in mammals has been well documented. This response requires the involvement of both the immune and the neuroendocrine systems, and stimulates the release of glucocorticoids and biogenic amines into the circulation via a hypothalamic-pituitary-adrenal gland axis (Axelrod and Reisine 1984; Bateman et al. 1989). In invertebrate immunocytes, the release of biogenic amines is observed after incubation of the hemolymph with CRH or adrenocorticotropic hormone (ACTH; Ottaviani et al. 1992, 1993a). Moreover, these immunocytes have been seen to contain immunoreactive molecules to a number of enzymes responsible for the synthesis of biogenic amines, such as tyrosine hydroxylase and dopamine β-hydroxylase (Ottaviani et al. 1993a). Overall, the results indicate that the stress response in invertebrates would appear to occur in a manner extremely similar to that in vertebrates, i.e. CRH → ACTH → biogenic amines. Together with this indirect but apparently main pathway of biogenic amine release, there seems to be a second, direct route induced by CRH and not mediated by ACTH (Ottaviani et al. 1992). It should be underlined that, while stress response in vertebrates requires the presence of numerous organs and cells, in invertebrates it is concentrated in a single immune neuroendocrine cell. In agreement with Blalock (1984) and Blalock and Smith (1985), we have proposed for the invertebrate immunocyte the term "mobile immune-brain" to describe the complex situation whereby mobile cells are able to recognize a multitude of stimuli and actuate a similarly complex response in which primitive mechanisms of both an immune and neuroendocrine type are intertwined (Ottaviani et al. 1991, 1993a).

Bearing in mind that mammal cytokines such as IL-1, IL-2, TNF-α and others are involved in stress response (Imura et al. 1991; Schöbitz et al. 1994), we have studied the involvement of these molecules in invertebrate stress response. The incubation of molluscan hemolymph with CRH provoked a release of biogenic amines higher than controls and IL-2 incubation, but lower than that observed following pre-incubation with IL-2. Moreover, no biogenic amine release was detected if the hemolymph was incubated with anti-IL-2 monoclonal antibody (mAb) alone or with CRH after pre-incubation with anti-IL-2 mAb. In order to clarify this behavior, we performed immunocytochemical and cytofluorimetric experiments. First, the immunocytes contained immunoreactive molecules to anti-IL-2 mAb and anti-CRH polyclonal antibody, and the IL-2 immunoreactivity disappeared after pre-absorption of the anti-IL-2 mAb not only with IL-2 but also with CRH. Flow cytometry confirmed these results, indicating the recognition by anti-IL-2 mAb of an epitope common both to IL-2 and CRH (Fig. 8) (Ottaviani et al. 1994). Further studies performed on biogenic amine release in the presence of other cytokines, i.e. IL-1α, and TNF-α, produced similar results to those obtained with IL-2 and CRH (Ottaviani et al. 1995a).

On the whole, these findings suggest the existence on the molluscan immunocyte membrane of an ancestral receptor capable of binding both IL-2 and CRH. Moreover, the competition between this neuropeptide and a large number of cytokines supports the idea that invertebrate cytokine receptors show a certain degree of promiscuity. In this respect, it should be stressed that the structure of mammalian cytokine receptors has been described as multi-subunits, with the same subunit shared by different cytokine receptors, such as IL-2, IL-4 and IL-7 (Taga and Kishimoto 1992; Kondo et al. 1993; Noguchi et al. 1993; Russell et al. 1993).

Intriguing findings have been obtained studying the relationship between growth factors and stress response. Indeed, it has been demonstrated that PDGF-AB and TGF-β1 are involved in different ways in biogenic amine release: the former inhibits and the latter stimulates this release (Table 6; Ottaviani et al. 1997a). The growth factor amine release is mediated by the CRH-ACTH-biogenic amine axis (Ottaviani et al. 1998). The addition of PDGF-AB to hemolymph pre-incubated with ACTH followed by CRH provoked a signif-

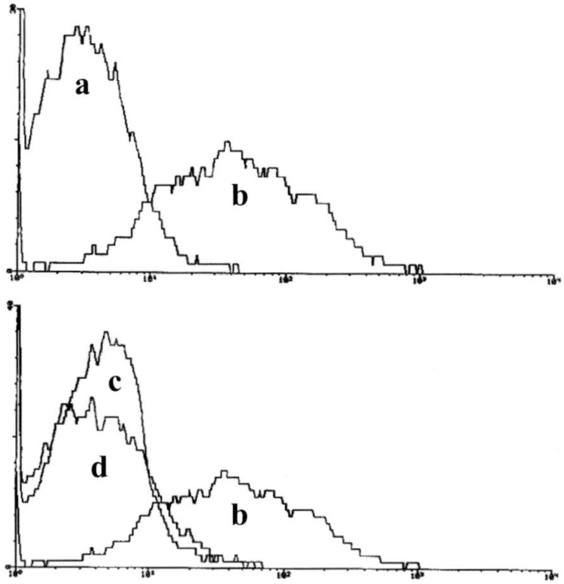

Fig. 8. Determination by flow cytometry of IL-2-like molecules on molluscan immunocyte plasma membrane by anti-IL-2 monoclonal antibody (*b*) and competition of this antibody with CRH (*c*) or IL-2 (*d*). *a* Control; *ordinate* cell number; *abscissa* fluorescence intensity. (Ottaviani et al. 1994, reprint with permission of the publisher)

Table 6. Concentrations of biogenic amines determined by HPLC in cell-free hemolymph (serum) of *Mytilus galloprovincialis* in the presence of 20 ng/ml PDGF-AB, 5 pg/ml TGF-β1, 10^{-8} M CRH and 10^{-8} M ACTH, respectively. (Modified from Ottaviani et al. 1997a)

	MS	PDGF-AB	TGF-β1	CRH	ACTH
NA	151.33 ± 5.69	107.33 ± 8.62*	275.33 ± 11.68*	265.67 ± 12.90*	229.00 ± 14.18*
A	33.03 ± 0.99	b.s.	50.10 ± 3.12*	47.64 ± 2.24*	43.40 ± 1.28*
DA	108.33 ± 6.10	118.20 ± 7.45	115.30 ± 7.35	107.85 ± 5.10	117.00 ± 5.00

The mean ± SD is reported. NA, Norepinephrine; A, epinephrine; DA, dopamine; b.s., below the sensitivity of the HPLC assay. The detection limit is 20 pg. Data are expressed as pg/ml. Statistical analysis was performed by ANOVA. *$P < 0.01$ vs. MS, Marine solution (control).

icant increase in biogenic amine release, indicating the annulment of the PDGF-AB inhibitory effect (Table 7). In contrast, when TGF-β1 was added, a further increase in amine release was observed (Table 8). These findings could be a useful indication in verifying if growth factors act in a similar way in the stress response of vertebrates.

On the whole, the multiple functions of cytokines found in invertebrates underline another characteristic of mammalian cytokines, i.e. their great pleiotropicity (Akira et al. 1990).

Table 7. Concentrations of biogenic amines determined by HPLC in cell-free hemolymph (serum) of *Mytilus galloprovincialis* following addition of 20 ng/ml PDGF-AB alone and after pre-incubation with 10^{-8} M ACTH followed by 10^{-8} M CRH. (Modified from Ottaviani et al. 1997a)

	PDGF-AB	ACTH-CRH-PDGF-AB
NA	107.33 ± 8.62	149.67 ± 9.50*
A	b.s.	53.46 ± 2.90*
DA	118.20 ± 7.45	104.33 ± 13.0

The mean ± SD is reported. NA, norepinephrine; A, epinephrine; DA, dopamine; b.s., below the sensitivity of the HPLC assay. The detection limit is 20 pg. Data are expressed as pg/ml. Statistical analysis was performed by ANOVA. *$P < 0.01$ vs. PDGF-AB.

Table 8. Concentrations of biogenic amines determined by HPLC in cell-free hemolymph (serum) of *Mytilus galloprovincialis* following addition of 5 pg/ml TGF-β1 alone and after pre-incubation with 10^{-8} M ACTH followed by 10^{-8} M CRH. (Modified from Ottaviani et al. 1997a)

	TGF-β1	ACTH-CRH-TGF-β1
NA	275.33 ± 11.68	365.33 ± 13.05*
A	50.10 ± 3.12	137.02 ± 4.62*
DA	115.30 ± 7.35	136.00 ± 9.16

The mean ± SD is reported. NA, norepinephrine; A, epinephrine; DA, dopamine; b.s., below the sensitivity of the HPLC assay. The detection limit is 20 pg. Data are expressed as pg/ml. Statistical analysis was performed by ANOVA. *$P < 0.01$ vs. TGF-β1.

4 Conclusions

The paper reviews the different functional aspects of cytokine-related molecules in invertebrates and shows a close overlapping with those found in vertebrates. The cytokine-like molecules are present in different tissues of various invertebrate species, affect immune functions such as cell motility, chemotaxis, phagocytosis and cytotoxicity, intervene in stress response, regulate programmed cell death, mediate wound repair, participate in the aging process and show pleiotropicity, functional redundancy and receptor promiscuity.

As previously reported, other molecules, including POMC-products, neuropeptides and NOS, are involved in basic and fundamental functions in cell survival (Ottaviani and Franceschi 1997), and these superimposed functions are also supported by molecular biology studies. Gene homology with vertebrate species has been found for the POMC gene in both the trematode *Schistosoma mansoni* (Duvaux-Miret and Capron 1991) and the leech *Theromyzon tessulatum* (Salzet et al. 1997), for the NOS gene in the insect *Rhodnius prolixus* (Yuda et al. 1996), and for the proCRH gene in the insect *Mamestra brassicae* immunocyte line (Malagoli et al. 2002). To date, very little molecular biology data has been reported for invertebrate cytokines, and the hypothesized missing correlation between invertebrate and vertebrate cytokine genes that is emerging (for review see Beschin et al. 2001) might represent a very peculiar strategy followed by Nature in the evolution of cytokines.

Acknowledgements. This work was supported by the following grants CNR, MIUR.

References

Aizawa K, Sato F (1963) Culture de tissus de ver a soie, *Bombyx mori*, dans un milieu sans hémolymphe. Ann Epiphyt 14:125

Akira S, Hirano T, Taga T, Kishimoto T (1990) Biology of multifunctional cytokines: IL 6 and related molecules (IL 1 and TNF). FASEB J 4:2860–2867

Axelrod J, Reisine TD (1984) Stress hormones: their interaction and regulation. Science 22:452–459

Ballarin L, Franchini A, Ottaviani E, Sabbadin A (2001) Morula cells as the major immunomodulatory hemocytes in ascidians: evidences from the colonial species *Botryllus schlosseri*. Biol Bull 201:59–64

Bateman A, Singh A, Kral T, Solomon S (1989) The immune-hypothalamic-pituitary-adrenal axis. Endocr Rev 10:92–112

Bates WR (1998) Evolutionary implications of FGF and distal-less expressions during proximal-distal axis formation in the ampulla of a direct-developing ascidian, *Molgula pacifica*. Biol Bull 194:241–243

Beck G, Habicht GS (1986) Isolation and characterization of a primitive interleukin-1-like protein from invertebrate, *Asterias forbesi*. Proc Natl Acad Sci USA 83:7429–7433

Beck G, Vasta R, Marchalonis J, Habicht GS (1989) Characterization of interleukin-1 activity in tunicates. Comp Biochem Physiol 92B:93–98

Beck G, O'Brien RF, Habicht GS, Stillman DL, Cooper EL, Raftos DA (1993) Invertebrate cytokines. III: invertebrate interleukin-1-like molecules stimulate phagocytosis by tunicate and echinoderm cells. Cell Immunol 146:284–299

Beck G, Habicht GS (1994) Invertebrate cytokines. Ann NY Acad Sci 712:206–212

Beck G, Cardinale S, Wang L, Reiner M, Sugumaran M (1996) Characterization of a defense complex consisting of interleukin 1 and phenol oxidase from the hemolymph of the tobacco hornworm, *Manduca sexta*. J Biol Chem 271:11035–11038

Beck G (1998) Macrokines: invertebrate cytokine-like molecules? Front Biosci 3:d559–d569

Berridge MJ (1987) Inositol triphosphate and diacylglycerol: two interacting second messengers. Annu Rev Biochem 56:159–193

Beschin A, Bilej M, Torreele E, De Baetselier P (2001) On the existence of cytokines in invertebrates. Cell Mol Life Sci 58:801–814

Blalock JE (1984) The immune system as a sensory organ. J Immunol 132:1067–1070

Blalock JE, Smith EM (1985) The immune system: our mobile brain? Immunol Today 6:115–117

Blalock JE (1994) The syntax of immune-neuroendocrine communication. Immunol Today 15: 504–511

Bryant PJ (1988) Localized cell death by mutations in a *Drosophila* gene coding for a transforming growth factor-β homolog. Dev Biol 128:386–395

Burdine RD, Chen EB, Kwok SF, Stern MJ (1997) *egl-17* encodes an invertebrate fibroblast growth factor family member required specifically for sex myoblast migration in *Caenorhabditis elegans*. Proc Natl Acad Sci USA 94:2433–2437

Burke RD, Watkins RF (1991) Stimulation of starfish coelomocytes by interleukin-1. Biochem Biophys Res Commun 180:579–584

Buttke TM, Sandstrom PA (1994) Oxidative stress as a mediator of apoptosis. Immunol Today 15:7–10

Carney DH, Scott DL, Gordon EA, LaBelle EF (1985) Phosphoinositides in mitogenesis: neomycin inhibits thrombin stimulated phosphoinositides turnover and inhibition of cell proliferation. Cell 42:479–488

Caselgrandi E, Kletsas D, Ottaviani E (2000) Neutral endopeptidase-24.11 (NEP) deactivates PDGF- and TGF-beta-induced cell shape changes in invertebrate immunocytes. Cell Biol Int 24:85–90

Casey ML, Smith JW, Nagai K, MacDonald PC (1993) Transforming growth factor-beta1 inhibits enkephalinase (EC 3.4.24.11) gene expression in human endometrial stromal cells and sex skin fibroblasts in culture. J Clin Endocrinol Metab 77:144–150

Cherbas L (1973) The induction of an injury reaction in cultured haemocytes from saturniid pupae. J Insect Physiol 19:2011–2023

Chjiwa T, Mishima A, Hagiwara M, Sano M, Hayashi K, Inoue T, Naito K, Tashioka T, Hidaka H (1990) Inhibition of forskolin-induced neurite outgrowth and protein phosphorylation by a newly synthesized selective inhibitor of cyclic AMP-dependent protein kinase, N-[2-(p-bromocinnamylamino)ethyl]-5-isoquinolinesulfonamide (H-89), of PC12D pheochromocytoma cells. J Biol Chem 265:5267–5272

Clark RAF (1996) Wound repair. Overview and general considerations. In: Clark RAF (ed) The molecular and cellular biology of wound repair. Plenum Press, New York, pp 3–50

Collins MK, Perkins GR, Rodriguez-Tarduchy G, Nieto MA, Lopez-Rivas A (1994) Growth factors as survival factors: regulation of apoptosis. BioEssays 16:133–138

Connelly JC, Chambless R, Holiday D, Chittenden K, Johnson AR (1993) Up-regulation of neutral endopeptidase (CALLA) in human neutrophils by granulocyte-macrophage colony-stimulating factor. J Leuk Biol 53:585–590

Connors V, DeBuron I, Granath W (1995) *Schistosoma mansoni*: interleukin-1 increases phagocytosis and superoxide production by hemocytes and decreases output of cercariae in schistosome-susceptible snail *Biomphalaria glabrata*. Exp Parasitol 80:139–148

Cooper EL, Franchini A, Ottaviani E (1995) Earthworm coelomocytes possess immunoreactive cytokines and POMC-derived peptides. Anim Biol 4:25–29

Cross M, Dexter TM (1991) Growth factors in developmental, transformation, and tumorigenesis. Cell 64:271–280

Darras S, Nishida H (2001) The BMP/CHORDIN antagonism controls sensory pigment cell specification and differentiation in the ascidian embryo. Dev Biol 236:271–288

Davidson B, Swalla BJ (2001) Isolation of genes involved in ascidian metamorphosis: epidermal growth factor signaling and metamorphic competence. Dev Genes Evol 211:190–194

Delikat SE, Galvani DW, Zuzel M (1994) A function of CD10 on bone marrow stroma. Br J Haematol 87:655–657

Di Virgilio F, Lew DP, Pozzan T (1984) Protein kinase C activation of physiological processes in human neutrophil at vanishingly small cytosolic Ca^{2+} levels. Nature 310:691–693

Duvaux-Miret O, Capron A (1991) Proopiomelanocortin gene and gene products: a comparative study. Adv Neuroimmunol 1:41–57

Fainzilber M, Smit AB, Syed NI, Wildering WC, Hermann PM, van der Schors RC, Jimenez C, Li KW, van Minnen J, Bulloch AGM, Ibanez CF, Garaerts WPM (1996) CRNF, a molluscan neurotrophic factor that interacts with the p75 neurotrophin receptor. Science 274:1540–1543

Franceschi C, Cossarizza A, Monti D, Ottaviani E (1991) Cytotoxicity and immunocyte markers in cells from the freshwater snail *Planorbarius corneus* (L) (Gastropoda, Pulmonata): implications for the evolution of natural killer cells. Eur J Immunol 21:489–493

Franceschi C, Monti D, Sansoni P, Cossarizza A (1995) The immunology of exceptional individuals: the lesson of centenarians. Immunol Today 16:12–16

Franchini A, Fontanili P, Ottaviani E (1995) Invertebrate immunocytes: relationship between phagocytosis and nitric oxide production. Comp Biochem Physiol 110B:403–407

Franchini A, Kletsas D, Ottaviani E (1996a) Immunocytochemical evidence of PDGF- and TGF-beta-like molecules in invertebrate and vertebrate immunocytes: an evolutionary approach. Histochem J 28:599–605

Franchini A, Miyan JA, Ottaviani E (1996b) Induction of ACTH- and TNF-alpha-like molecules in the hemocytes of *Calliphora vomitoria* (Insecta, Diptera). Tissue Cell 28:587–592

Franchini A, Ottaviani E (2000) Repair of molluscan tissue injury: role of PDGF and TGF-beta. Tissue Cell 32:312–321

Franchini A, Peruzzi E, Ottaviani E (2003) Morphochemical age-related changes in the nematode *Caenorhabditis elegans*: immunoperoxidase localization of cytokine- and growth factor-like molecules. Eur J Histochem 47:75–80

Gomez-Escobar N, Lewis E, Maizels RM (1998) A novel member of the transforming growth factor-beta (TGF-β) superfamily from the filarian nematodes *Brugia malayi* and *B. pahandi*. Exp Parasitol 88:200–209

Govindarajan V, Ramachandran RK, George JM, Shakes DC, Tomlinson CR (1995) An ECM-bound, PDGF-like growth factor and a TGF-alpha-like growth factor are required for gastrulation and spiculogenesis in the *Lytechinus* embryo. Dev Biol 172:541–551

Granath WO Jr, Connors VA, Tarleton RL (1994) Interleukin 1 activity in haemolymph from strains of the snail *Biomphalaria glabrata* varying in susceptibility to the human blood fluke, *Schistosoma mansoni*: presence, differential expression, and biological function. Cytokine 6: 21–27

Grotendorst GR, Martin GR, Pencev D, Sodek J, Harvey AK (1985) Stimulation of granulation tissue formation by platelet-derived growth factor in normal and diabetic rats. J Clin Invest 176:2323–2329

Harris CE, Staley D, Conn PM (1985) Diacylglycerols and protein kinase C. Potential amplifying mechanism for Ca^{2+}-mediated gonadotropin-releasing hormone-stimulated luteinizing hormone release. Mol Pharmacol 27:532–536

Hayashi I, Perez-Magallanes M, Rossi JM (1992) Neurotrophic factor-like activity in *Drosophila*. Biochem Biophys Res Commun 184: 73–79

Hermann PM, van Kesteren RE, Wildering WC, Painter SD, Reno JM, Smith JS, Kumar SB, Garaerts WPM, Ericsson LH, Smit AB, Bulloch AGM, Nagle GT (2000) Neurotrophic actions of a novel molluscan epidermal growth factor. J Neurosci 20:6355–6364

Hill RJ, Sternberg PW (1992) The gene *lin-3* encodes an inductive signal for vulval development in *C. elegans*. Nature 358:470–476

Holgate ST, Lewis RA, Austin KF (1980) Role of adenylate cyclase in immunologic release of mediators from rat mast cells: agonist and antagonist effects of purine- and ribose-modified adenosine analogs. Proc Natl Acad Sci USA 77:6800–6804

Hug H, Sarre TF (1993) Protein kinase C: divergence in signal transduction? Biochem J 291: 329–343

Hughes TK Jr, Smith EM, Chin R, Cadet P, Sinisterra J, Leung MK, Shipp MA, Scharrer B, Stefano GB (1990) Interaction of immunoreactive monokines (interleukin 1 and tumor necrosis factor) in the bivalve mollusc *Mytilus edulis*. Proc Natl Acad Sci USA 87:4426–4429

Hughes TK Jr, Smith EM, Barnett JA, Charles R, Stefano GB (1991) LPS stimulated invertebrate hemocytes: a role for immunoreactive TNF and IL-1. Dev Comp Immunol 15:117–122

Hughes TK Jr, Smith EM, Leung MK, Stefano GB (1992) Immunoreactive cytokines in *Mytilus edulis* nervous and immune interactions. Acta Biol Hung 43:269–273

Imura H, Fukata J, Mori T (1991) Cytokines and endocrine function: an interaction between the immune and neuroendocrine systems. Clin Endocrinol (Oxf) 35:107–115

Kawamura K, Shibata T, Saget O, Peel D, Bryant PJ (1999) A new family of growth factors produced by fat body and active on *Drosophila* imaginal disc cells. Development 126:211–219

Kelley MR, Kidd S, Deutsch WA, Young MW (1987) Mutations altering the structure of epidermal growth factor-like coding sequences at the *Drosophila Notch* locus. Cell 51:539–548

Khouri RK, Koudsi B, Deune G, Hong SP, Ozbek MR, Serdar CM, Song S-Z, Pierce GF (1993) Tissue generation with growth factors. Surgery 114:374–380

Kimura A, Iwamoto I, Nakagawa N, Tomoka H, Yoshida S (1991) Neutral endopeptidase (EC 3.4.24.11) does not hydrolyze recombinant human interleukin-1beta. Immunol Lett 28: 109–114

Kletsas D, Sassi D, Franchini A, Ottaviani E (1998) PDGF and TGF-beta induce cell shape changes in invertebrate immunocytes via specific cell surface receptors. Eur J Cell Biol 75:362–366

Kondepudi A, Johnson A (1993) Cytokines increase neutral endopeptidase activity in lung fibroblasts. Am J Respir Cell Mol Biol 8:43–49

Kondo M, Takeshita T, Ishii N, Nakamura M, Watanabe S, Arai K-I, Sugamura K (1993) Sharing of the interleukin-2 (IL-2) receptor gamma chain between receptors for IL-2 and IL-4. Science 262:1874–1877

Kopczynski CC, Arlon AK, Fechtel K, Kooh PJ, Muskavitch MA (1988) *Delta*, a *Drosophila* neurogenic gene, is transcriptionally complex and encodes a protein related to blood coagulation factors and epidermal growth factor of vertebrates. Genes Dev 2:1723–1735

Kresse H, Schonherr E (2001) Proteoglycans of the extracellular matrix and growth control. J Cell Physiol 189:266–274

Kutty G, Kutty K, Samuel W, Duncan T, Jaworski C, Wiggert B (1998) Identification of a new member of transforming growth factor-beta superfamily in *Drosophila*: the first *activin* gene. Biochem Biophys Res Commun 246:644–649

Lelong C, Mathieu M, Fevrel P (2000) Structure and expression of mGDF, a new member of the transforming growth factor-beta superfamily in the bivalve mollusc *Crassostrea gigas*. Eur J Biochem 267:3986–3993

Malagoli D, Mandrioll M, Ottaviani E (2003) Cloning and characterisation of a procorticotrophin-releasing homone in the IZD-MB-0503 immunocyte line from the insect *Mamestra brassicae*. Peptides 23:1829–1836

Mayer M, Noble M (1994) N-acetyl-L-cysteine is a pluripotent protector against cell death and enhancer of trophic factor-mediated cell survival *in vitro*. Proc Natl Acad Sci USA 91: 7496–7500

Miya T, Morita K, Ueno N, Satoh N (1996) An ascidian homologue of vertebrate BMPs-5–8 is expressed in the midline of the anterior neuroectoderm and in the midline of the ventral epidermis of the embryo. Mech Dev 57:181–190

Mustoe TA, Pierce GF, Thomason A, Gramates P, Sporn MB, Deuel TF (1987) Accelerated healing of incisional wounds in rats induced by transforming growth factor-beta. Science 237: 1333–1336

Nakayama A, Satou Y, Satoh N (2001) Isolation and characterization of genes that are expressed during *Ciona intestinalis* metamorphosis. Dev Genes Evol 211:184–189

Neuman-Silberberg FS, Schupbach T (1993) The *Drosophila* dorsoventral patterning gene *gurken* produces a dorsally localized RNA and encodes a TGF-alpha-like protein. Cell 74:165–174

Nishikawa T, Sasano H, Omura M, Suematsu S (1996) Regulation of expression of the steroidogenic acute regulatory (StAR) protein by ACTH in bovine adrenal fasciculata cells. Biochem Biophys Res Commun 223:12–18

Nisticò G (1993) Communications among central nervous system, neuroendocrine and immune systems: interleukin-2. Prog Neurobiol 40:463–475

Noguchi M, Nakamura Y, Russell SM, Ziegler SF, Tsang M, Cao X, Leonard WJ (1993) Interleukin-2 receptor gamma chain: a functional component of the interleukin-7 receptor. Science 262: 1877–1880

Ottaviani E, Cossarizza A. Ortolani C, Monti D, Franceschi C (1991) ACTH-like molecules in gastropod molluscs: a possible role in ancestral immune response and stress. Proc R Soc Lond B 245:215–218

Ottaviani E, Caselgrandi E, Petraglia F, Franceschi C (1992) Stress response in the freshwater snail *Planorbarius corneus* (L.) (Gastropoda, Pulmonata): interaction between CRF, ACTH and biogenic amines. Gen Comp Endocrinol 87:354–360

Ottaviani E, Caselgrandi E, Franchini A, Franceschi C (1993a) CRF provokes the release of norepinephrine by hemocytes of *Viviparus ater* (Gastropoda, Prosobranchia): further evidence in favour of the evolutionary hypothesis of the "mobile immune-brain". Biochem Biophys Res Commun 193:446–452

Ottaviani E, Franchini A, Franceschi C (1993b) Presence of several cytokine-like molecules in molluscan hemocytes. Biochem Biophys Res Commun 195:984–988

Ottaviani E, Paemen LR, Cadet P, Stefano GB (1993c) Evidence for nitric oxide production and utilization as a bacteriocidal agent by invertebrate immunocytes. Eur J Pharmacol Environ Toxicol Pharmacol 248:319–324

Ottaviani E, Franchini A, Caselgrandi E, Cossarizza A, Franceschi C (1994) Relationship between corticotropin-releasing factor and interleukin-2: evolutionary evidence. FEBS Lett 351:19–21

Ottaviani E, Caselgrandi E, Franceschi C (1995a) Cytokines and evolution: *in vitro* effects of IL-1alpha, IL-1beta, TNF-alpha and TNF-beta on an ancestral type of stress response. Biochem Biophys Res Commun 207:288–292

Ottaviani E, Franchini A, Cassanelli S, Genedani S (1995b) Cytokines and molluscan immune responses. Biol Cell 85:87–91

Ottaviani E, Franceschi F (1997) The invertebrate phagocytic immunocyte: clues to a common evolution of immune and neuroendocrine systems. Immunol Today 18:169–174

Ottaviani E, Caselgrandi E, Kletsas D (1997a) Effect of PDGF and TGF-beta on the release of biogenic amines from invertebrate immunocytes and their possible role in the stress response. FEBS Lett 403:236–238

Ottaviani E, Franchini A, Kletsas D, Bernardi M, Genedani S (1997b) Involvement of PDGF and TGF-beta in cell migration and phagocytosis in invertebrate and human immunocytes. Anim Biol 6:95–99

Ottaviani E, Sassi D, Kletsas D (1997c) PDGF- and TGF-beta-induced changes in cell shape of invertebrate immunocytes: effect of calcium entry blockers. Eur J Cell Biol 74:336–347

Ottaviani E, Caselgrandi E, Kletsas D (1998) The CRH-ACTH-biogenic amine axis in invertebrate immunocytes activated by PDGF and TGF-beta. FEBS Lett 427:255–258

Ottaviani E, Malagoli D, Kletsas D (1999) Platelet-derived growth factor and transforming growth factor-beta induce shape changes in invertebrate immunocytes via multiple signalling pathways and provoke the expression of Fos-, Jun- and SMAD-family members. Comp Biochem Physiol 122B:389–395

Ottaviani E, Barbieri D, Franchini A, Kletsas D (2000a) PDGF and TGF-beta partially prevent 2-deoxy-D-ribose-induced apoptosis in the fat body cell line IPLB-LdFB from the insect *Lymantria dispar*. J Insect Physiol 46:81–87

Ottaviani E, Franchini A, Malagoli D, Genedani S (2000b) Immunomodulation by recombinant human interleukin-8 and and its signal transduction pathways in invertebrate hemocytes. Cell Mol Life Sci 57:506–513

Ottaviani E, Barbieri D, Malagoli D, Kletsas D (2001) Involvement of PI 3-kinase, PKA and PKC in PDGF- and TGF-beta-mediated prevention of 2-deoxy-D-ribose-induced apoptosis in the insect cell line, IPLB-LdFB. Cell Biol Int 25:171–177

Ouwe-Missi-Oukem-Boyer O, Porchet E, Capron A, Dissous C (1994) Characterization of immunoreactive TNF-alpha molecules in the gastropod *Biomphalaria glabrata*. Dev Comp Immunol 18:211–218

Padgett RW, Johnson RD, Gelbart WM (1987) A transcript from a *Drosophila* pattern gene predicts a protein homologous to the transforming growth factor-beta family. Nature 325: 81–84

Paeman LR, Porchet-Hennere E, Masson M, Leung MK, Hughes TK Jr, Stefano GB (1992) Glial localization of interleukin-1alpha in invertebrate ganglia. Cell Mol Neurobiol 12:463–471

Pierart ME, Najdovski T, Appelboom TE, Deschodt-Lamckmann M (1988) Effect of human endopeptidase 24.11 ("enkephalinase") on IL-1-induced thymocyte proliferation activity. J Immunol 140:3808–3811

Pierce GF, Mustoe TA, Lingelbach J, Masakowski VR, Griffin GL, Senior RM, Duel TF (1989) Platelet-derived growth factor and transforming growth factor-beta enhance tissue repair activities by unique mechanisms. J Cell Biol 109:429–440

Pierce GF, Tarpley JE, Yanagihara D, Mustoe TA, Fox GM, Thomason A (1992) Platelet-derived growth factor (BB homodimer), and transforming growth factor-beta 1, and basic fibroblast growth factor in dermal wound healing. Neovessel and matrix formation and cessation of repair. Am J Pathol 140:1375–1388

Ponce MR, Micol JL, Peterson KJ, Davidson EH (1999) Molecular characterization and phylogenetic analysis of SpBMP5-7, a new member of the TGF-beta superfamily expressed in sea urchin embryos. Mol Biol Evol 16:634–645

Prendergast RA, Suzuki M (1970) Invertebrate protein simulating mediators of delayed hypersensitivity. Nature 227:277–279

Prendergast RA, Lutty GA, Scott AL (1983) Direct inflammation: the phylogeny of lymphokines. Dev Comp Immunol 7:629–632

Quaglino D, Nanney LB, Ditesheim JA, Davidson JM (1991) Transforming growth factor-beta stimulates wound healing and modulates extracellular matrix gene expression in pig skin. Incisional wound model. J Invest Dermatol 97:34–42

Raftos DA, Cooper EL, Habicht GS, Beck G (1991) Invertebrate cytokines: tunicate cell proliferation stimulated by an interleukin 1-like molecule. Proc Natl Acad Sci USA 88:9518–9522

Ratcliffe NA, Rowley AF, Fitzgerald SW, Rhodes CP (1985) Invertebrate immunity: basic concepts and recent advances. Int Rev Cytol 97:183–354

Ren P, Lim C-S, Johnsen R, Albert PS, Pilgrim D, Riddle DL (1996) Control of *C. elegans* larval development by neuronal expression of a TGF-beta homolog. Science 274:1389–1391

Roberts AB, Sporn MB, Assoian RK, Smith JM, Roche NS, Wakefield LM, Heine UI, Liotta LA, Falange V, Kehrl JH, Fauci AS (1986) Transforming growth factor type beta: rapid induction of fibrosis and angiogenesis in vivo and stimulation of collagen formation *in vitro*. Proc Natl Acad Sci USA 83:4167–4171

Roubin R, Vatcher G, Popovici C, Coulier F, Thierry-Mieg J, Pantarotti P, Birnbaum D, Baillie D, Thierry-Mieg D (1999) *let-756*, a *C. elegans* fgf essential for worm development. Oncogene 18: 6741–6747

Russell SM, Keegan AD, Harada N, Nakamura Y, Noguchi M, Leland P, Friedmann MC, Miyajima A, Puri RK, Paul WE, Leonard WJ (1993) Interleukin-2 receptor gamma chain: a functional component of the interleukin-4 receptor. Science 262:1880–1883

Salzet M, Salzet-Raveillon B, Cocquerelle C, Verger-Bocquet M, Pryor SC, Rialas CM, Laurent V, Stefano GB (1997) Leech immunocytes contain proopiomelanocortin: nitric oxide mediates hemolymph proopiomelanocortin processing. J Immunol 159:5400–5411

Schöbitz B, De Kloet ER, Holsboer F (1994) Gene expression and function of interleukin-1, interleukin-6 and tumor necrosis factor in the brain. Prog Neurobiol 44:397–432

Staal FYT, Stephen WE, Roederer M, Anderson MT, Herzenberg LA (1992) Glutathione deficiency and human immunodeficiency virus infection. Lancet 339:909–912

Stenzel P, Angerer LM, Smith BJ, Angerer RC, Vale WW (1994) The *univin* gene encodes a member of the transforming growth factor-beta superfamily with restricted expression in the sea urchin embryo. Dev Biol 166:149–158

Stefano GB, Smith EM, Hughes TK Jr (1991) Opioid induction of immunoreactive interleukin-1 in *Mytilus edulis* and human immunocytes: an interleukin-1-like substance in invertebrate neural tissue. J Neuroimmunol 32:29–34

Suzuki Y, Yandell MD, Roy PJ, Krishna S, Savage-Dunn C, Ross RM, Padgett RW, Wood WB (1999) A BMP homolog acts as a dose-dependent regulator of body size and male tail patterning in *Caenorhabditis elegans*. Development 126:241–250

Taga T, Kishimoto T (1992) Cytokine receptors and signal transduction. FASEB J 6:3387–3396

Tamaoki T (1991) Use and specificity of staurosporine, UCN-01, and calphostin C as protein kinase inhibitors. Method Enzymol 201:340–347

Tharaux P-L, Stefanski A, Ledoux S, Soleilhac J-M, Ardaillou R, Dussaule J-C (1997) EGF and TGF-beta regulate neutral endopeptidase expression in renal vascular smooth muscle cells. Am J Physiol 272:C1836–C1843

Van der Velden VH, Naber BA, van der Spoel P, Hoogseden HG, Versnel MA (1998) Cytokines and glucocorticoids modulate human bronchila epithelial cell peptidase. Cytokine 10:55–65

Vaughn JL, Louloudes EJ (1978) Isolation of two promoting fractions from insect hemolymph. In Vitro 14:351

Wharton KA, Johansen KM, Xu T, Artavanis-Tsakonas S (1985) Nucleotide sequence from the neurogenic locus *Notch* implies a gene product that shares homology with proteins containing EGF-like repeats. Cell 43:567–581

Wielgus JJ, Caldwell GA, Nichols RL, White CF (1990) Purification, properties, and titer of a hemolymph trophic factor in larvae and pupae of *Manduca sexta*. Insect Biochem 20:65–72

Wieprecht M, Wieder T, Geilen CC (1994) N-[2-Bromocinnamyl(amino)ethyl]-5-isoquinoline-sulphonamide (H89) inhibits incorporation of choline into phosphatidylcholine via inhibition of choline kinase and has no effect on the phosphorylation of CTP: phosphocholine cytidylytransferase. Biochem J 297:241–247

Williams CM, Kambysellis MP (1969) *In vitro* action of ecdysone. Proc Natl Acad Sci USA 63:231

Wittwer D, Franchini A, Ottaviani E, Wiesner A (1999) Presence of IL-1- and TNF-like molecules in *Galleria mellonella* (Lepidoptera) haemocytes and in an insect cell line from *Estigmene acraea* (Lepidoptera). Cytokine 11:637–642

Yamasu K, Watanabe H, Kohchi C, Soma G, Mizuno D, Akasaka K, Shimada H, Suyemitsu T, Ishihara K (1995) Molecular cloning of a cDNA that encodes the precursor to several exogastrula-inducing peptides, epidermal-growth-factor-related polypeptides of the sea urchin *Anthocidaris crassispina*. Eur J Biochem 228:515–523

Yuda M, Hirai M, Miura K, Matsumura H, Ando K, Chinzei Y (1996) cDNA cloning, expression and characterization of nitric oxide synthase from the salivary glands of the blood-sucking insect *Rhodnius prolixus*. Eur J Biochem 242:807–812

Cytokines in *Drosophila* Hematopoiesis and Cellular Immunity

D.L. Bodian,[2] S. Leung,[1] H. Chiu,[2] and S. Govind[1]

Abstract. Hematopoiesis is a complex, multistep process in which progenitor cells undergo distinct cellular changes of proliferation and differentiation to give rise to mature blood cells in circulation. Many of the genetic and molecular events that drive these changes have been characterized in mammals, frogs, and zebra fish, and more recently in the insect model system *Drosophila melanogaster*. Blood cells in *Drosophila* are actively involved in fighting infections and the cellular immune responses are intimately tied to the process of hematopoiesis. In this article, we briefly review the fundamental similarities in *Drosophila* and mammalian hematopoiesis and highlight the potential roles of four cytokines/growth factors in *Drosophila* hematopoiesis and cellular immunity.

1 Introduction

For more than 80 years, *Drosophila* has served as a powerful model system for studying a variety of basic developmental cellular and molecular processes. The availability of the *Drosophila* genome sequence and a curated set of predicted genes has greatly enhanced its power as an experimental tool not only for the study of human disease (Rubin et al. 2000) but also as a model insect for vector biology and pest control (Schneider 2000). With the knowledge of a majority of the gene products encoded in the *Drosophila* genome, it is now possible to systematically evaluate the functions and functional relationships of all members of a gene family or subfamily.

As initiators of molecular events from outside the cell, cytokines and growth factors play a critical role in cell proliferation and differentiation. The role of mammalian cytokines and growth factors in normal hematopoiesis and leukemia has been the subject of intense study (Lotem and Sachs 2002). In contrast, studies on cytokines in general and their roles in hematopoiesis in particular in lower vertebrates and invertebrates have been relatively limited. Most of our information on their role in blood formation comes from *Drosophila* where a number of cytokines and growth factors were already identified on the basis of genetic studies prior to the availability of the genome

[1]Department of Biology J526, City College and the Graduate Center of the City University of New York, 138th Street and Convent Avenue, New York, New York 10031, USA
[2]Department of Functional Genomics, Novartis Pharmaceuticals Corp, 556 Morris Avenue, Summit, New Jersey 07901, USA

Progress in Molecular and Subcellular Biology
A. Beschin (Ed.) Invertebrate Cytokines

sequence. These proteins, like their mammalian counterparts, regulate cell signaling and physiology and misregulation of their functions leads to aberrant development or disease. In addition to the characterized cytokines, more than 60 additional cytokine/hormones/growth factor-like proteins are predicted in the *Drosophila* genome (FlyBase). In this chapter, we discuss both the known and potential roles of cytokines in *Drosophila* hematopoiesis.

2
Blood Cell Formation and Function in *Drosophila*

2.1
Hematopoiesis and Blood Cells in *Drosophila*

In vertebrates, hematopoiesis occurs in a variety of different locations (yolk sac, aorta, gonad/mesonephros region, liver, spleen, thymus, bone marrow), depending on the life cycle stage of the animal. Blood cell formation during the embryonic stages is referred to as the "primitive" phase of hematopoiesis, whereas blood cells that are formed after birth and continue through the adult stages come from the "definitive" phase of hematopoiesis. In mouse and humans, there are at least eight different lineages of blood cells that fall into two broad categories, the myeloid and the lymphoid, each derived from distinct progenitor cell types. The lymphoid progenitors give rise to B and T cells that engage in adaptive immune responses. The myeloid progenitors give rise to macrophages. These cells are phagocytic and play a role in the innate immune defense response.

Vertebrate hematopoietic cells utilize many growth factors, chemokines and cytokines to regulate their survival, proliferation and differentiation (Geijsen et al. 2001; Mackay 2001; Zon 2001). In the absence of any infection, the bone marrow and the stromal cells are the primary source of hematopoietic cytokines. After infection, cytokines are produced by activated macrophages and T helper cells.

Like vertebrates, hematopoiesis in simple invertebrates such as insects also occurs in distinct temporal phases. In *Drosophila*, the first hematopoietic phase occurs during embryonic stages and a second phase is observed during larval stages. There are two mature blood cell types in the *Drosophila* embryo and four cell types in the larva. Like their vertebrate counterparts, these cells also engage in cellular immune reactions. However, a clear role for the involvement of cytokines and growth factors in *Drosophila* hematopoiesis and immunity is only beginning to emerge.

2.1.1
Embryonic Blood Cells in Drosophila

Tissues that support the production of phagocytic blood cells (macrophage-like) in the early and mid-stage embryo have been identified in the head mesoderm (HM, Fig. 1A; Rugendorff et al. 1994). Embryonic macrophages consume apoptotic cells generated during normal development (Franc et al. 1999). During mid-embryogenesis, a cluster of a second set of cells (crystal cell precursors, CCP, Fig. 1B) serves as precursors of the second blood cell type, the

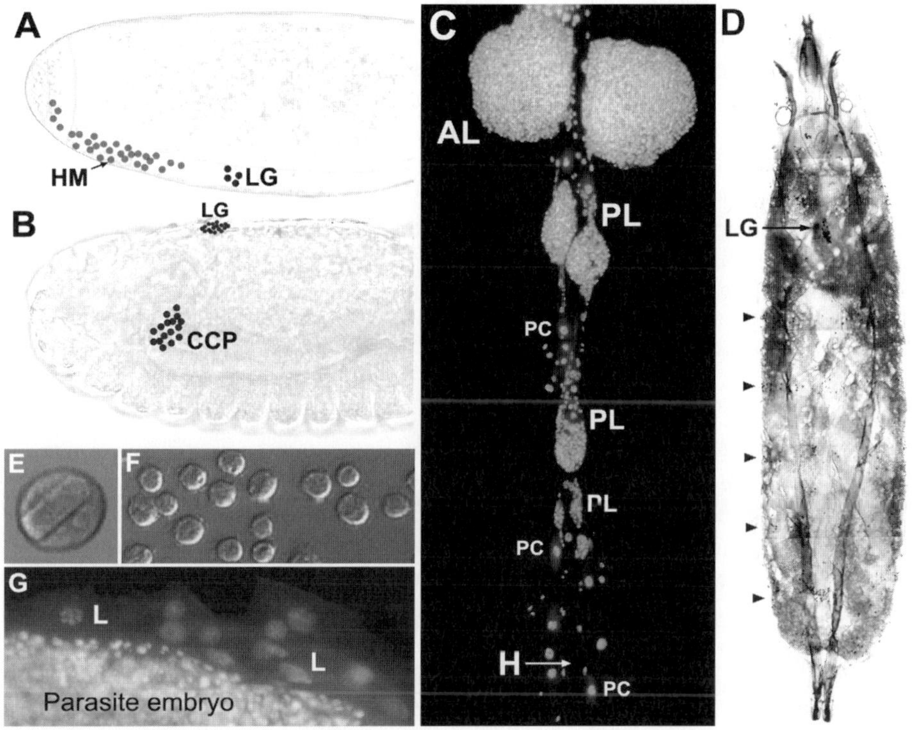

Fig. 1. Hematopoietic regions and blood cells in the *Drosophila* embryo and larva. **A** Schematic representation of an early and **B** mid-stage embryo showing the locations of regions where blood cells originate. Cells in the anterior or head mesoderm (*HM*) and crystal cell precursors (*CCP*) each give rise to the embryonic blood cells. A third population of blood cells (*LG*) gives rise to the embryonic lymph gland, which becomes functional during larval stages. **C** Photograph of a larval lymph gland dissected from a wild type third instar larva. Notice presence of pairs of lobes around the dorsal vessel. Anterior lobes (*AL*), posterior lobes (*PL*), pericardial cells (*PC*) and heart (*H*). **D** Crystal cells present in sessile islets underneath the cuticle, segmentally localized (indicated by *arrowheads*), visualized after treating larvae at high temperature. **E** Crystal cells and **F** plasmatocytes in circulation. **G** Lamellocytes wrapped around an egg of a wasp embryo (also see Fig. 2F)

crystal cells (Lebestky et al. 2000). The biological function of crystal cells in the embryo is obscure.

2.1.2
Blood Cells in the Larva: Three Compartments

During embryogenesis, the future larval hematopoietic organ called the lymph gland is formed from clusters of lateral mesodermal cells (Rugendorff et al. 1994; Tepass et al. 1994). These cells assemble on the dorsal side of the embryo at the anterior end of the dorsal vessel (Fig. 1B). Much like a vertebrate capillary, the dorsal vessel is a simple, tubular structure. It is made up of cardioblasts and pericardial cells (Rugendorff et al. 1994).

The embryonic lymph gland is thought to be non-functional. After hatching, the larval lymph gland lobes grow in size. By the third instar stage, the lymph gland is immune competent and consists of four to six pairs of lobes that are organized serially around the dorsal vessel (Fig. 1C). The anterior-most pair of lobes (AL) is the largest in size whereas the lobe size becomes progressively smaller in the posterior regions. The lymph gland lobes harbor precursors of mature blood cells in circulation. The narrower portion of the dorsal vessel on the anterior side is called the aorta, whereas the wider portion of the dorsal vessel in the posterior segments is referred to as the heart (Fig. 1C, H; Rizki 1978). The heart directs the flow of hemolymph in the body cavity. The entire structure is held in place beneath the epidermis by alary muscles.

The primary blood cell type in circulation is called the plasmatocyte. Like the cells in the embryo, they are phagocytic; in addition to phagocytosis of apoptotic cells (Figs. 1F, 2A), these cells can also consume bacteria and viruses. Crystal cells are also present in circulation, although they are not as numerous (less than 5% of all blood cells in circulation) as the plasmatocytes. Crystal cells (Fig. 1E) contain the enzymes and substrates for the melanization reactions that accompany cellular immune responses. The cellular precursors of plasmatocytes and crystal cells have been identified in the larval lymph gland lobes in transmission electron microscopy studies (Shrestha and Gateff 1982; Lanot et al. 2001). These lobes additionally contain specialized secretory cells, defined as such on the basis of morphology. Secretory cells are apparently not released into circulation and their function in the lymph gland is not known (Lanot et al. 2001).

The remaining population of larval blood cells is sessile and is located in small groups underneath the integument. These "sessile islets" are organized segmentally (Fig. 1D). Mature plasmatocytes and crystal cells are found in roughly the same proportion in these islets as they are in circulation.

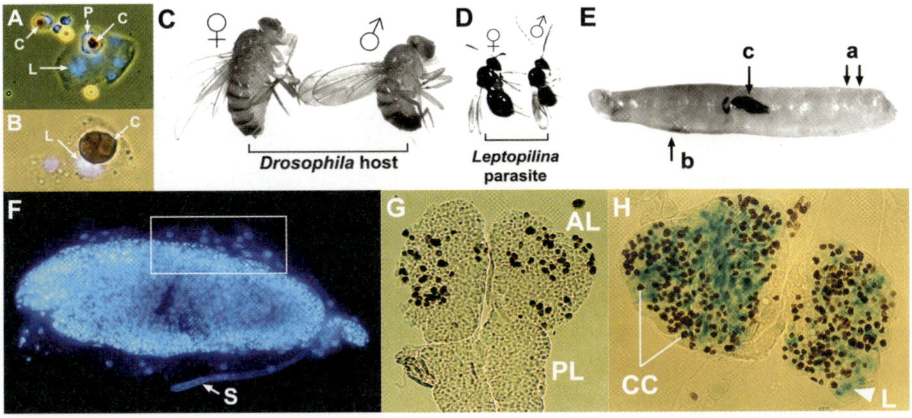

Fig. 2. Cellular immune reactions in the *Drosophila* larva. (**A, B**) A smear of hemolymph from larva carrying the *Black cells* mutation that causes the death of crystal cells. Dead crystal cells provoke cellular immune responses: melanization within crystal cells themselves (labeled *C* in **A** and **B**), phagocytosis or encapsulation of melanized crystal cells by plasmatocytes (*P*) or lamellocytes (*L*, **A** and **B**). Plasmatocyte nucleus is stained with a fluorescent dye. **C, D** Adult *Drosophila* and parasitic wasps, *Leptopilina boulard,* photographed together. **E** A third instar larva carrying a melanized parasite wasp egg visible through its cuticle. *Arrows* show the following positions: (*a*) common site of infection, (*b*) location of the lymph gland and (*c*) melanized capsule. **F** A capsule of lamellocytes around a parasite embryo, stalk, *S*. **G, H** Anterior lobes of the larval lymph glands dissected from uninfected (**G**), or wasp infected (**H**), third instar larvae carrying the *Bc*/+ and *l(3)msn*^{03349}/+ markers for the crystal cells and lamellocytes respectively. Note that lamellocytes (*blue staining*, *L*) are visible only upon wasp infection response after infection (**H**). There is also a drastic increase in the number of crystal cells after wasp infection (compare dark cells, labeled *CC* in **G** and **H**)

2.2
Cellular Immune Reaction in the Larva: Encapsulation of Parasite Eggs

Like blood cells in higher organisms, blood cells in *Drosophila* carry out a number of important life-sustaining functions. These include (1) maintenance of the extracellular matrix, (2) clotting and wound healing, (3) phagocytosis of microbes and cellular debris (Fig. 2A, B) and (4) encapsulation of macroscopic pathogens. The encapsulation reaction involves division and differentiation of blood cell precursors and appearance of the activated cell type called lamellocyte. The encapsulation reaction is considered in more detail here (also see Fig. 2).[3]

[3]When *Drosophila* larvae or adults are infected by bacteria or fungi, a powerful anti-microbial response is activated. Anti-microbial peptide genes are strongly upregulated in the fat body. This transcriptional activation is under the control of the NF-κB pathways (Hoffmann et al. 1999; Khush et al. 2001). This humoral response is not activated when *Drosophila* larvae are infected by wasps (Nicolas et al. 1996).

Second instar or early-third instar larvae of *Drosophila melanogaster* serve as hosts for several parasitoid wasp species (e.g. Melk and Govind 1999). Female wasps inject eggs into the larval hemocoel. This activates the cellular immune response of the host. The egg is sequestered and killed within a cellular capsule made up mostly of lamellocytes (Fig. 2F), although some plasmatocytes are also present. The capsule darkens (Fig. 2E), presumably due to a phenoloxidase cascade involving enzymes localized in the crystal cells. This enzymatic cascade leads to the formation of melanin. The site of egg injection (generally the posterior most segments) and location of mature capsule (generally mid-abdominal region) varies (Fig. 2E).

There is good evidence that plasmatocytes, lamellocytes and crystal cell precursors within the lymph gland exhibit specific responses to wasp parasitization. (1) Lamellocytes that make up the capsule are not normally present in unparasitized larvae but differentiate in large numbers in the lymph gland in response to wasp parasitization (Lanot et al. 2001; Sorrentino et al. 2002) (Fig. 2G, H). (2) The number of mature crystal cells in the lymph gland significantly increases in parasitized third instar hosts (Sorrentino et al. 2002) (Fig. 2G, H). (3) Lymph glands from parasitized early-third instar hosts exhibit better than twofold increase in mitotic index prior to the dispersal of the cells in the anterior lobes (Sorrentino et al. 2002). Consistent with these changes in the lymph gland, a modest increase in blood cell concentration (~ two fold) that occurs after parasitization has been recorded (Rizki and Rizki 1992).

The process of encapsulation involves (1) recognition of the egg after its deposition at position "a" (Fig. 2E), (2) a set of highly-controlled hematopoietic changes involving division and differentiation of precursors in the lymph gland (position "b", Fig. 2E), (3) migration of mature immune cells from the lymph gland to the wasp egg in the hemocoel and (4) formation of the capsule (typically at position "c" in Fig. 2E). Because these positions span the length of the larval body, it is likely that one or more long-range signals, involving response to guidance cues are required for the successful completion of this process. In addition to the precursor cells, the roles of the mature cells in the innate immune response have also been studied. There is some evidence that crystal cells are unlikely to play an important role in recognition of the foreign invader while plasmatocytes could contribute to this process. Mutant larvae lacking crystal cells readily encapsulate parasite eggs without subsequent melanization of the wasp egg (Rizki and Rizki 1990).

3
Molecular Conservation in the Regulation of *Drosophila* and Mammalian Hematopoiesis

3.1
Transcription Factors and Lineage Specification

The two embryonic cell types, the macrophage and the crystal cells arise from the same precursor cells but undergo stepwise changes of commitment and differentiation to form the two distinct lineages. Mutations in specific *Drosophila* genes (*serpent, glial cells missing, lozenge, ushaped*) have been identified that disrupt the specification of all blood cells or either one of the two individual lineages. Based on their loss-of-function phenotypes, a role for these genes in specific hematopoietic steps have been identified (for details see Fig. 3). *serpent* (*srp*) appears to be required for plasmatocyte differentiation and its

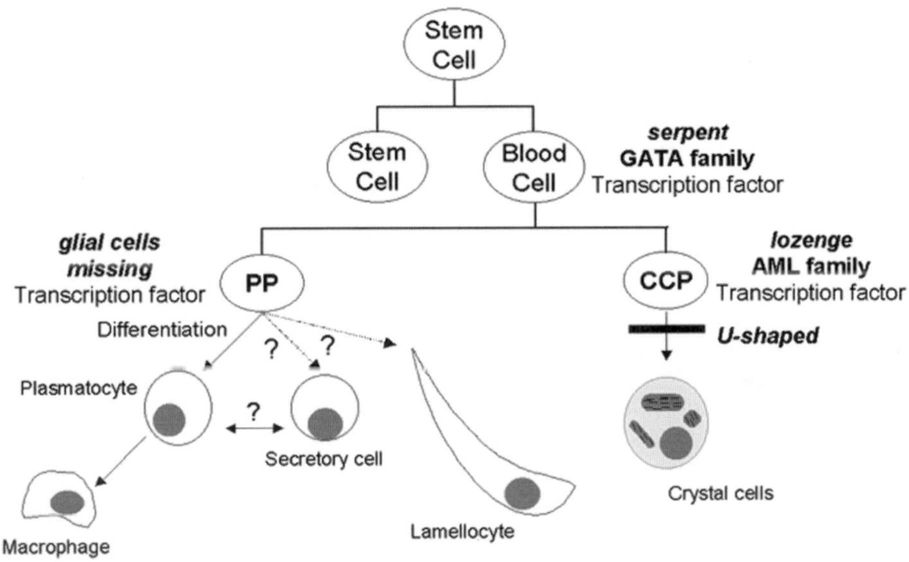

Fig. 3. A model for hematopoiesis in *Drosophila* involving transcription factors and signaling pathways. The GATA-family transcription factor Serpent (*Srp*) is required for the specification of all blood cells as *srp* null embryos exhibit a reduction in the number of blood cell precursors and complete loss of both blood lineages (Rehorn et al. 1996; Lebestky et al. 2000). Glial Cells Missing (*Gcm*) is necessary for the differentiation of the macrophage lineage in the embryo (Bernardoni et al. 1997; Lebestky et al. 2000), whereas the Runt-family transcription factor Lozenge (*Lz*) is essential for the specification of the crystal cell population (Lebestky et al. 2000). Friend-of-GATA (FOG, also called Ushaped or Ush) is a multitype zinc-finger protein that acts as a negative regulator of crystal cell development. Mutant *ushaped* embryos show an increase in the number of crystal cells whereas forced expression of Ush protein in crystal cell precursors reduces crystal cell number in the embryo (Fossett et al. 2001). *PP* Plasmatocyte Precursor; *CCP* crystal cell precursor

expression is maintained in the plasmatocyte lineage throughout embryogenesis (Rehorn et al. 1996). Furthermore, expression of *lozenge* (similar to human *AML-1, 2, 3* and mouse *CBFA-1, 2, 3*) and *ushaped* (similar to the mammalian Friend of GATA, *FOG*) requires Serpent protein function (Lebestky et al. 2000; Fossett et al. 2001). Thus, much as in mammalian hematopoiesis, a cascade of transcription factors acting to promote and repress cell differentiation controls hematopoietic lineage specification in the *Drosophila* embryo, and likely also during larval stages. Genetic analysis has shown that mouse homologues of these *Drosophila* genes are central to mammalian hematopoiesis: knockout mutations of *GATA1, 2, 3, FOG-2* and *CBFA2* genes in mice eliminate either both primitive and definitive hematopoiesis or alter specific aspects of defined lineages (reviewed in Cantor and Orkin 2001; Fossett and Schulz 2001; Meister and Govind 2003).

3.2 Signal Transduction Pathways in Blood Cell Division, Differentiation and Migration

Almost all of the major signal transduction pathways that control cell division and differentiation in mammals are also found in *Drosophila* (Rubin et al. 2000). Components of these pathways perform the same biochemical functions and appear in the same functional order in fly and mammalian cells. These pathways are activated by cytokines or growth factors. To date, components in four such pathways (PDGF/VEGF, Toll/NF-κB, JAK/STAT and steroid hormone) have been implicated in aspects of *Drosophila* embryonic or larval hematopoiesis. In this section, we briefly review the evidence for their participation in hematopoiesis and discuss the role of *Drosophila* cytokines and growth factors involved, as well as similarities in structure and function between the fly and human homologues. It is important to bear in mind that each of these pathways also controls several other biological functions in the fly and their non-hematopoietic functions are not included in our discussion.

3.2.1 Pvfs

The PDGF/VEGF (platelet derived growth factor/vascular endothelial growth factor) family includes a number of cytokines that are potent regulators of vasculogenesis and angiogenesis in mammals. The bioactive forms of PDGFs and VEGFs are dimers which, like other dimeric extracellular signaling molecules, bind to specific cell surface receptors, inducing dimerization and triggering downstream signaling events. Proteins are classified in this family based on the presence of the VEGF/PDGF homology domain, a domain of approximately 100 amino acids characterized by eight conserved cysteines. Crystal structures of mammalian family members (Oefner et al. 1992; Muller et al. 1997) reveal that the VEGF/PDGF homology domain forms a cysteine knot structure held together by three pairs of disulfide bonds between six of the eight conserved cysteines (*orange*, Fig. 6A). The remaining two conserved cysteines (*green*, Fig. 6A) form intermolecular disulfide bonds between the two subunits of the dimer.

Sequence analysis identified three *Drosophila* proteins with the VEGF/PDGF domain, Pvf1, Pvf2, and Pvf3 (Pvf for PDGF/VEGF family) (Duchek et al. 2001; Heino et al. 2001; Cho et al. 2002). All eight cysteine residues are conserved in the *Drosophila* homologues except for the second cysteine in Pvf2 (Duchek et al. 2001; Cho et al. 2002) and all three Pvfs are predicted to form cysteine knot structures similar to those of mammalian PDGFs and VEGFs.[4] Future studies will establish whether the *Drosophila* Pvfs dimerize, whether they form homo- or heterodimers, and if their oligomerization is functionally important.

Human PDGF/VEGF proteins are synthesized as precursors which require proteolytic processing for full activity. The processing cleaves domains amino- or carboxy-terminal to the VEGF/PDGF homology domain which are thought to regulate localization, compartmentalization, and/or receptor binding activities of the proteins (Ferrara 1999; Li et al. 2000; Olofsson et al. 1999). For example, in addition to amino-terminal propeptides, VEGF-C and -D have carboxy-terminal cysteine-rich regions which are proteolytically processed to generate active species (Joukov et al. 1997; Stacker et al. 1999). The *Drosophila* Pvf proteins may undergo a similar mechanism of regulation since they contain carboxy-terminal cysteine-rich sequences reminiscent of that of VEGF-C (Heino et al. 2001; Cho et al. 2002).

The biologically active forms of mammalian PDGFs and VEGFs transduce signals by binding to specific cell surface receptors. This signaling is characterized by overlapping specificity of the ligands and redundancy in function. In *Drosophila*, the same theme of overlapping specificity and redundancy is maintained. While three Pvfs have been identified, there is only a single VEGF receptor-like protein (DmVEGFR, Heino et al. 2001; PVR, Duchek et al. 2001; VEGFR, Cho et al. 2002). Like the mammalian VEGF receptors, the *Drosophila* VEGF receptor is a tyrosine kinase. It is expressed in embryonic blood cells throughout their migrations (Heino et al. 2001; Cho et al. 2002).

The Pvfs and the VEGF receptors have been functionally implicated in migration of the blood cells in the embryo (Cho et al. 2002). The evidence is as follows: (1) Mutations in *Vegfr* lead to a disruption of migration of blood

[4]There is some question in the literature as to whether Pvf2 is a PDGF/VEGF family member since it lacks one of the apparently critical residues for this family. The missing cysteine corresponds to one of the residues forming the interchain disulfide bonds and there is evidence from studies on VEGFs -C and -D and PDGF-B that interchain disulfide bonds are not essential for conformationally stable, active dimers (Prestrelski et al. 1994; Joukov et al. 1997; Stacker et al. 1999).

cells and the site of migratory arrest correlates with the expression of the Pvfs. (2) While the inactivation of individual *Pvf* genes had no effect on blood cell migration, simultaneous inactivation of all three Pvf genes via RNAi resulted in the same phenotype as *Vegfr* inactivation. (3) Ectopic expression of any one of the ligands re-routes blood cell migration (Cho et al. 2002).

There is some evidence that the RAS-MAP kinase pathway is activated during blood cell migration. Immunostaining experiments with an anti-diphospho-MAPK antibody revealed the presence of this activated kinase in migrating blood cells. Furthermore, expression of a dominant negative RAS protein caused an arrest in blood cell migration that is similar to that observed in the *Vegfr* mutants. Together, these observations in the *Drosophila* embryo are significant, since in vertebrates the VEGF pathways are associated with blood vessel and lymphatic development and it is believed that the ancestral function of the VEGF pathways was to guide the movement of blood cells (Cho et al. 2002).

3.2.2
The Toll/NF-κB Cassette

Components of the Toll/I-κB/NF-κB pathway (Fig. 4) were originally identified as maternal-effect mutations, whose functions are required for the specification of the dorsoventral axis of the early embryo (Govind 1999). The intracellular components of this pathway include the Toll receptor, cytoplasmic proteins Tube and Pelle, inhibitory protein I-κB or Cactus and NF-κB protein, Dorsal.

Evidence for a role of the Toll/NF-κB pathway components in *Drosophila* hematopoiesis comes from analysis of mutants or transgenic strains. Upregulation of the Toll/Dorsal segment of this cascade by dominant gain-of-function alleles of *Toll*, strong recessive loss-of-function alleles of *cactus*, or overexpression of Dorsal protein or mammalian p50, all produce hematopoietic defects (Gerttula et al. 1988; Roth et al. 1991; Govind 1996; Qiu et al. 1998). These defects include changes in blood cells that also accompany egg encapsulation, the difference being that hematopoietic defects caused by mutations are much stronger than those observed after wasp infection. Thus, in hematopoietic mutants, there is (1) abnormally high concentration of circulating blood cells including large numbers of lamellocytes, (2) lymph gland hypertrophy and early dispersal, and (3) the appearance of melanized encapsulations of self-tissue. These defects appear to be due to the result of misregulation of hemocyte proliferation and/or differentiation programs and lead to the death of the animal. Consistently, loss-of-function of *cactus* also increases mitosis in the lymph gland and in circulation, while downregulation of the pathway due to loss-of-function in *Toll, tube* and *pelle* has the opposite effect: circulating blood cell concentrations are significantly reduced (Qiu et al. 1998). These observations thus strongly suggest that the Toll/Tube/Pelle/Cactus

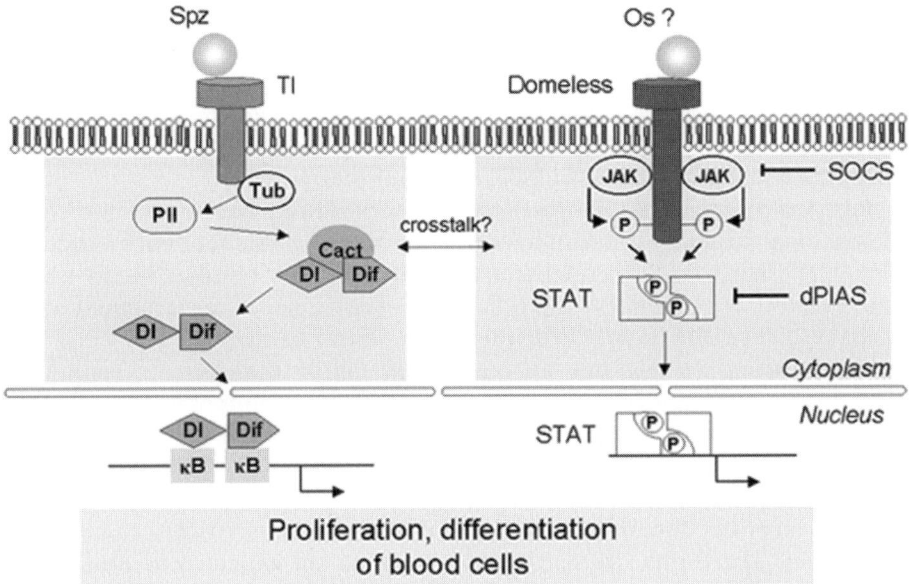

Fig. 4. Schemes showing components of the NF-κB (*left*) and JAK/STAT (*right*) pathways of *Drosophila* proposed to be active in the larval lymph gland. The Toll receptor in the NF-κB pathway activates cytoplasmic proteins Tube (*Tub*) and Pelle (*Pll*), which in turn leads to the activation of Rel-family proteins Dorsal (*Dl*) or *Dif*. Cactus is the I-κB inhibitor of the pathway and keeps the Dorsal/Dif proteins from nuclear localization in the absence of an extracellular signal. In other places in the larval life cycle, Toll is activated by Spätzle in the extracellular compartment. The Domeless receptor of the JAK/STAT pathway of *Drosophila* is activated by Unpaired (*Upd*)/Outstretched (*os*) (or potentially by Outstretched-like proteins). The signal is transduced to the JAK kinase Hopscotch. Phosphorylation of the transcription factor STAT results in its dimerization and nuclear localization. Upregulation of either the Toll/Cactus or JAK/STAT pathways results in hematopoietic disorders of overproliferation and differentiation of lymph glands

cascade plays a direct role in the regulation of larval hemocyte density. Interestingly, mammalian homologues of members of the *Drosophila* Toll/I-κB pathway have been identified, and are also involved in hematopoiesis and innate immunity (for review Meister and Govind 2003).

The cytokine Spätzle (spz) presumably activates Toll, localized uniformly to the plasma membrane of the embryo. The Spätzle protein is synthesized in the embryo, secreted into the extracellular perivitelline space where it is proteolytically processed, and is thought to activate Toll only in the ventral and ventrolateral regions of the embryo (Stein and Nusslein-Volhard 1992; Morisato and Anderson 1994; DeLotto and DeLotto 1998). The carboxy-terminal 106 amino acid residues of Spätzle are sufficient for activation of Toll (Morisato and Anderson 1994).

Spätzle is predicted to be a member of the cysteine knot superfamily based on the similarity in the spacing of cysteine residues clustered in its carboxy-terminal active domain with that of cysteines in known cysteine knot proteins (Morisato and Anderson 1994; DeLotto and DeLotto 1998; Mizuguchi et al. 1998). This carboxy-terminal active fragment exists as a disulfide-linked homodimer, shown to be biologically active (DeLotto and DeLotto 1998). Multiple, alternatively spliced forms of Spätzle have been found, most of which affect the propeptide sequence (DeLotto et al. 2001).

Proteins belonging to the cysteine knot superfamily have conserved three-dimensional structure and can have little or no detectable sequence similarity. Within the superfamily, proteins are subdivided into families based on the conservation of sequence. Unlike the Pvf proteins, which are assigned to the PDGF/VEGF family based on sequence similarity to mammalian VEGF and PDGF proteins, Spätzle has no recognized family members in mammals. However, bioinformatics analyses of the *Drosophila* genome sequence and the derived predicted proteins have suggested that additional fly homologues of Spz exist (Parker et al. 2001). The predicted proteins CG14928, CG18318, CG9972 and CG7104/CG14533 have significant sequence similarity to the carboxy-terminal domain of Spätzle. CG9196 was also predicted to belong to this family. These putative Spätzle-like proteins are expected to function as specific ligands of the many *Drosophila* Toll-like receptors (Tauzig et al. 2000; Parker et al. 2001).

There is some evidence that Spätzle plays a role in hematopoiesis. Like loss-of-function mutations in *Toll*, *tube* and *pelle*, loss-of-function mutations in *spätzle* (spz^{rm7}) results in lowered blood cell concentration during larval stages. Furthermore, when subjected to parasite infections, these mutants are unable to defend themselves as vigorously as their heterozygous siblings (Sorrentino and Govind, unpubl.). This latter immune-deficiency phenotype is also observed in $Toll^{-}$, $tube^{-}$ and $pelle^{-}$ mutants (Sorrentino and Govind, unpubl.). These results support the idea that activation of Toll by active Spätzle results in transduction of the intracellular signal through Tube and Pelle proteins to maintain wild-type levels of blood cell concentrations. Lowered blood cell counts via mutations in any one of these genes result in compromising the immune response of encapsulation.

3.2.3 *The JAK/STAT Pathway*

The JAK/STAT pathways in mammals are activated by a variety of cytokines and growth factors and regulate, among other biological functions, different aspects of hematopoiesis (Ward et al. 2000; Rane and Reddy 2002). Upregulation of the *Drosophila* Hopscotch/Stat92E pathway (Fig. 4), much like the Toll/Cactus pathway, also activates larval hematopoiesis. Hopscotch (Hop), a receptor-associated tyrosine kinase, is a member of the JAK family of protein tyrosine kinases. Dominant gain-of-function alleles $hop^{Tumorous\text{-}lethal}$ ($hop^{Tum\text{-}l}$ and hop^{T42}) cause constitutive activation of the downstream signaling cascade, producing hematopoietic defects similar to those caused by the upregulation of the Toll/Cactus cassette. *hop*$^{Tum\text{-}l}$/+ adults that also carry loss-of-function alleles of *stat* exhibit a reduced frequency of melanized encapsulation of self tissue (Harrison et al. 1995; Luo et al. 1995). Conversely, loss-of-function of *Dpias* results in an increased likelihood of capsule formation in adults, while overexpression of *dpias* has an opposite effect (Betz et al. 2001). Dpias, or *Drosophila* PIAS (protein inhibitors of activated STAT) is an inhibitor of JAK/STAT pathway and is thought to negatively regulate *stat92E*.

To date, only a single extracellular ligand for the JAK/STAT pathway has been reported. Unpaired (*upd*), also referred to as Outstretched (*os*), is a secreted protein that activates the JAK/STAT pathway in *Drosophila* presumably by binding to the transmembrane receptor Mom/Domeless, the Hop-associated receptor in embryogenesis (Harrison et al. 1998; Brown et al. 2001; Chen et al. 2002). Because Domeless has some characteristics of mammalian cytokine receptors, particularly those of the gp130 family, it was proposed that Unpaired may belong to the same superfamily as the known ligands for these receptors, the superfamily of 4-helical bundle cytokines (Brown et al. 2001; Chen et al. 2002). The lack of a detectable sequence relationship of Unpaired to these ligands does not preclude membership in this superfamily since, like cysteine knot proteins, members of the superfamily can have low sequence identity despite a common three-dimensional fold (Mott and Campbell 1995). Further work is required to determine whether upd is indeed an IL-6-like cytokine.

Despite the lack of sequence similarity to known proteins, Unpaired appears to have homologues of unknown function encoded within the *Drosophila* genome. A search of the proteins predicted from the *Drosophila* genome sequence (Adams et al. 2000) revealed at least two putative family members (Fig. 5). All three genes are located on the X chromosome at band 17A. The putative protein CG5988 shows 30% sequence identity to Upd over 285 aligned

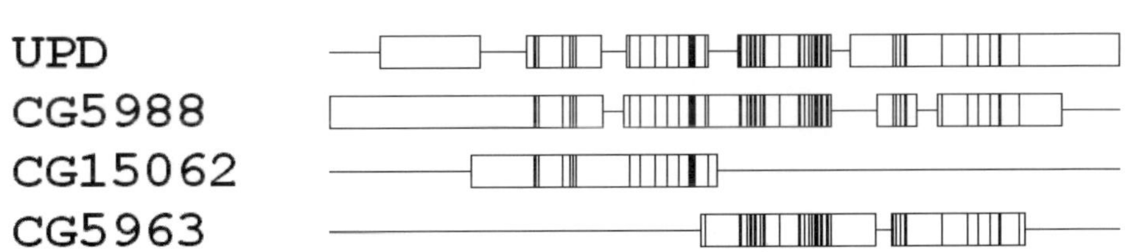

Fig. 5. Multiple sequence alignment of Unpaired and Unpaired-like protein sequences. The sequences are drawn as open rectangles, with gaps inserted solely for the purpose of alignment represented by horizontal connecting bars. Residues conserved in at least three of the sequences are indicated by black vertical bars, with the width of the bars proportional to the number of consecutive conserved residues. The sequences of CG5988, CG15062, and CG5963 are translations of in silico gene predictions (Adams et al. 2000) and have not been confirmed experimentally. The alignment was generated using CLUSTALW (Thompson et al. 1994), manually edited, and displayed with GENEDOC. (Nicholas et al. 1997)

amino acids. Two additional predicted proteins, CG15062 and CG5963, have significant sequence similarity to the amino- and carboxy-terminal halves of Upd, respectively. Since these two predicted genes are adjacent in the genome and they have the same transcriptional orientation when aligned with the genomic sequence, it is likely that CG15062 and CG5963 represent the 5′ and 3′ ends of a single gene that is a third Upd-family member. Since the sequences of both Upd-like proteins are derived from ab initio gene predictions, experimental verification of the gene structures and sequences are required, as is confirmation that they represent expressed genes.

Although there is evidence for the JAK/STAT pathway in regulating immune responses, no clear role for Unpaired in *Drosophila* hematopoiesis has been reported. Overexpression of Unpaired protein in larvae does not result in hematopoietic defects as might be expected upon upregulation of the pathway (Morey et al. 2001). Further, loss-of-function mutations in *upd* do not suppress the hematopoietic defects observed in the *hop*$^{Tum\text{-}l}$ background nor do they compromise the immune capacity to fight wasp infections (Sorrentino and Govind, unpubl.). Given that Unpaired may not play a role in hematopoiesis or immunity, or may have redundant functions with other family members, a potential role of the other two Unpaired-like proteins in the activation of the JAK/STAT pathway in blood cells merits attention.

4 Other Pathways and Molecules in *Drosophila* Hematopoiesis

4.1 A Role for Ecdysone in Blood Cell Development and Differentiation

In addition to cytokines, steroid hormones also control the development and functions of blood cells in significant ways. Ecdysone, a steroid hormone, binds to and activates transcription factor receptors EcR and Ultraspiracle to orchestrate a complex developmental program that converts a larva to a fly. Ecdysone is synthesized in the ring gland, a neuroendocrine structure present in close proximity of the larval brain and the lymph gland.

Hormones control the ability of the immune responsive fat body in the induction of antimicrobial peptides (Meister and Richards 1996), and lymph gland in the encapsulation of the wasp egg (Sorrentino et al. 2002). In the latter case, developmental competence is acquired in the third larval instar, and animals at the previous instars are unable to respond to wasp infection. Indeed the lymph glands of first and second instar larvae are much smaller in size. Developmentally arrested, early-third instar larvae, with reduced ecdysone titers (as a result of a temperature-sensitive *ecdysoneless* mutation) fail to mount a robust encapsulation reaction in response to wasp infections. In these larvae, the frequency of lymph gland dispersal is strongly reduced: there is no increase in mitotic activity of lymph gland hemocytes in early-third instar wasp-parasitized mutant larvae; and residual lamellocytes normally present in the hemolymph of wasp-parasitized larvae are not present in circulation (Sorrentino et al. 2002). Such observations strongly suggest that lymph gland hemocytes gain immune competency, the ability of precursors to divide and differentiate under the influence of ecdysone.

A second and distinct requirement for ecdysone is observed at metamorphosis when the cells of the lymph gland undergo clear morphological changes: most of the lymph gland cells become active macrophages and invade larval structures to consume larval tissues undergoing histolysis (Lanot et al. 2001). In older, developmentally delayed mutant animals, lymph gland cells do not undergo this division to differentiation switch and become quite large due to continued proliferation (Sorrentino et al. 2002).

4.2 Tumor Necrosis Factor-like Protein

Cytokines belonging to the TNF (tumor necrosis factor) superfamily (TNFSF) have pleiotropic roles in the morphogenesis of various tissues. Most mammalian members of this family are expressed in the immune system, where they regulate cell proliferation, differentiation, activation, and apoptosis. The TNFSF proteins share a domain of approximately 150 amino acids called the TNF homology domain (THD). Sequence identity between family members is 20–30% and is limited to the region of this domain. Despite the low sequence identity between TNFSF members, crystal structures revealed that they have a highly conserved three-dimensional structure (Hymowitz et al. 2000) (Fig. 6B). Most TNFSF members are synthesized as type II transmembrane proteins, with the carboxy-terminal end, including the TNF homology domain, extracellular. Soluble forms of some of these proteins are generated by proteolysis and both the soluble and transmembrane forms of these ligands are active as trimers. All TNFSF members are capable of binding to receptors belonging to the TNF receptor superfamily (TNFRSF), transmembrane receptors that contain a homologous cysteine-rich domain. Recognition between the TNFSF and TNFRSF proteins occurs through their hallmark domains, the TNF homology domain of the ligands and the cysteine-rich domain of the receptors. Ligand binding induces trimerization of the receptors, which initiates downstream signaling events. Many downstream events that lead to the activation of caspases and transcription factors such as AP-1 and NF-κB have been characterized (Baud and Karin 2001).

A *Drosophila* gene *eiger* has recently been reported as encoding a member of the TNFSF. This gene corresponds to CG12919 and is proposed to serve as a ligand for the JNK pathway in control of apoptosis (Igaki et al. 2002). Another predicted protein with a cysteine-rich domain, CG6531, is proposed to be a member of the TNFRSF but not an orthologue of human TNF receptor (Adams

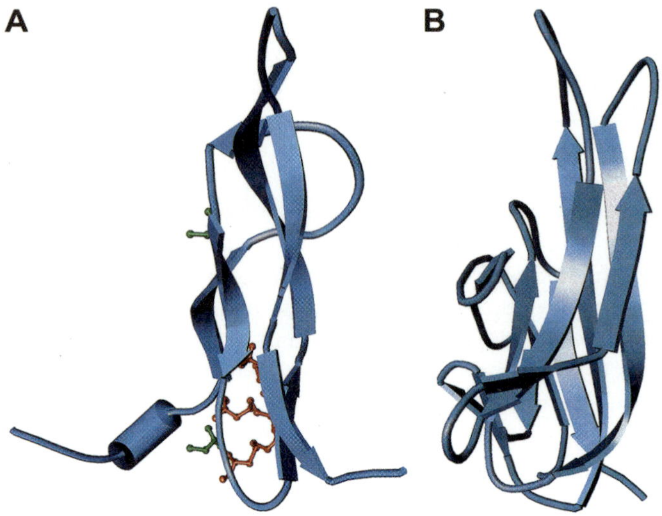

Fig. 6. Structure of representative PDGF/VEGF and TNF superfamily members. β-sheets and α-helices are drawn as arrows and cylinders, respectively. **A** The monomer of human VEGF (Muller et al. 1997). The conserved cysteines are displayed as ball and stick figures. Six cysteines (*orange*) form the three disulfide bonds of the cysteine knot fold and the two remaining cysteines (*green*) form disulfide bonds with partners in the second polypeptide chain of the homodimer. **B** The monomer of human TRAIL, a TNF superfamily member, is composed of antiparallel β-strands in a jellyroll fold (Hymowitz et al. 1999). The monomers associate to form symmetric homotrimers, with most of the residues conserved in the family responsible for the trimeric structure (Locksley et al. 2001). The *Drosophila* Pvfs and TNF-like protein are predicted to have folds similar to those of their human homologues. The monomers of Spätzle and its homologues are also predicted to have cysteine knot folds. (Morisato and Anderson 1994; Parker et al. 2001)

et al. 2000; Aravind et al. 2001). Because CG12919 is homologous to mammalian TNF ligands and CG6531 is similar to their known receptors, it is tempting to speculate that these two *Drosophila* proteins interact with each other. Experimental studies are required to test this hypothesis and to determine their possible roles in the control of hematopoiesis.

5 Concluding Remarks

There has been a recent surge of interest in studying *Drosophila* hematopoiesis and our understanding of the organization and development of hematopoietic structures and the molecules that regulate the production and function of blood cells is therefore growing. Some of the key players that control lineage specification and the pathways that regulate blood cell division and differentiation have been identified. In this context, the potential role of a few cytokines/cytokine-like molecules (Pvfs, Spätzle, Unpaired) has come to the fore and it is reasonable to assume that some of these molecules (or their *Drosophila* homologues) selectively activate pathways regulating blood cell migration, proliferation, or differentiation of hematopoietic tissues. The functions of other well-characterized cytokines/growth factors (Hedgehog, Decapentaplegic, Wingless, etc) remain to be explored.

The availability of whole genome sequences and high level of structural and functional similarities in the molecular components of signaling pathways in *Drosophila* and mammals will facilitate our understanding of how cytokines/growth factors have evolved relative to components in the intracellular compartments. Conservation of important cellular and developmental mechanisms across species (Rubin et al. 2000) underlies our confidence that knowledge of the cellular and molecular basis of hematopoiesis and immunity in *Drosophila* will lead to the rapid identification of human homologues and aid in the development of novel approaches to treating infections and injury in humans.

Acknowledgements. SG acknowledges funding from the American Cancer Society (RPG 98-228-01-DDC), American Heart Association Heritage Affiliate, Inc., NIH-RCMI RR03060-16 and PSC-CUNY. We thank RP Sorrentino for help with the Figures and colleagues for suggestions on the manuscript.

References

Adams MD, Celniker SE, Holt RA et al (2000) The genome sequence of *Drosophila melanogaster*. Science 287:2185–2195

Aravind L, Dixit VM, Koonin EV (2001) Apoptotic molecular machinery: vastly increased complexity in vertebrates revealed by genome comparisons. Science 291:1279–1284

Baud V, Karin M (2001) Signal transduction by tumor necrosis factor and its relatives. Trends Cell Biol 11:372–377

Bernardoni R, Vivancos B, Giangrande A (1997) *glide/gcm* is expressed and required in the scavenger cell lineage. Dev Biol 191:118–130

Betz A, Lampen N, Martinek S, Young M, Darnell JE (2001) A *Drosophila* PIAS homologue negatively regulates stat92E. Proc Natl Acad Sci USA 98:9563–9568

Brown S, Hu N, Hombria JC (2001) Identification of the first invertebrate interleukin JAK/STAT receptor, the *Drosophila* gene *domeless*. Curr Biol 11:1700–1705

Cantor AB, Orkin SH (2001) Hematopoietic development: a balancing act. Curr Opin Genet Dev 11: 513–519

Chen H-W, Chen X, Oh S-W, Marinissen MJ, Gutkind JS, Hou SX (2002) *mom* identifies a receptor for the *Drosophila* JAK/STAT signal transduction pathway and encodes a protein distantly related to the mammalian cytokine receptor family. Genes Dev 16:388–398

Cho NK, Keyes L, Johnson E, Heller J, Ryner L, Karim F, Krasnow MA (2002) Developmental control of blood cell migration by the *Drosophila* VEGF pathway. Cell 108:865–876

DeLotto Y, DeLotto R (1998) Proteolytic processing of the *Drosophila* Spätzle protein by Easter generates a dimeric NGF-like molecule with ventralising activity. Mech Dev 72:141–148

DeLotto Y, Smith C, DeLotto R (2001) Multiple isoforms of the *Drosophila* Spätzle protein are encoded by alternatively spliced maternal mRNAs in the precellular blastoderm embryo. Mol Gen Genet 264:643–652

Duchek P, Somogyi K, Jekely G, Beccari S, Rorth P (2001) Guidance of cell migration by the *Drosophila* PDGF/VEGF receptor. Cell 107:17–26

Ferrara N (1999) Molecular and biological properties of vascular endothelial growth factor. J Mol Med 77:527–543

Fossett N, Schulz RA (2001) Functional conservation of hematopoietic factors in *Drosophila* and vertebrates. Differentiation 69:83–90

Fossett N, Tevosian SG, Gajewshi K, Zhang Q, Orkin S, Schulz RA (2001) The friend of GATA proteins U-shaped, FOG-1 and FOG-2 function as negative regulators of blood, heart and eye development in *Drosophila*. Proc Natl Acad Sci USA 98:7342–7347

Franc NC, Heitzler P, Ezekowitz RA, White K (1999) Requirement for *croquemort* in phagocytosis of apoptotic cells in *Drosophila*. Science 284:1991–1994

Geijsen N, Koenderman L, Coffer PJ (2001) Specificity in cytokine signal transduction: lessons learned from the IL-3/IL-5/GM-CSF receptor family. Cytokine Growth Factor Rev 12:19–25

Gerttula S, Jin YS, Anderson KV (1988) Zygotic expression and activity of the *Drosophila* Toll gene, a gene required maternally for embryonic dorsal-ventral pattern formation. Genetics 119:123–133

Govind S (1996) Rel signalling pathway and the melanotic tumour phenotype of *Drosophila*. Biochem Soc Trans 24:39–44

Govind S (1999) Control of development and immunity by Rel transcription factors in *Drosophila*. Oncogene 18:6875–6887

Harrison DA, Binari R, Nahreini TS, Gilman M, Perrimon N (1995) Activation of a *Drosophila* Janus kinase (JAK) causes hematopoietic neoplasia and developmental defects. EMBO J 14: 2857–2865

Harrison DA, McCoon PE, Binari R, Gilman M, Perrimon N (1998) *Drosophila* unpaired encodes a secreted protein that activates the JAK signaling pathway. Genes Dev 12:3252–3263

Heino TI, Karpanen T, Wahlstrom G, Pulkkinen M, Eriksson U, Alitalo K, Roos C (2001) The *Drosophila* VEGF receptor homolog is expressed in hemocytes. Mech Dev 109:69–77

Hoffmann JA, Kafatos FC, Janeway CA, Ezekowitz RA (1999) Phylogenetic perspectives in innate immunity. Science 284:1313–1318

Hymowitz SG, Christinger HW, Fuh G, Ultsch M, O'Connell M, Kelley RF, Ashkenazi A, de Vos AM (1999) Triggering cell death: the crystal structure of Apo2L/TRAIL in a complex with Death Receptor 5. Mol Cell 4:563–571

Hymowitz SG, O'Connell MP, Ultsch MH, Hurst A, Totpal K, Ashkenazi A, de Vos AM, Kelley RF (2000) A unique zinc-binding site revealed by a high-resolution X-ray structure of homotrimeric Apo2L/TRAIL. Biochemistry 39:633–640

Igaki T, Kanda H, Yamamoto-Goto Y, Kanuka H, Kuranaga E, Aigaki T, Muira M (2002) Eiger, a TNF superfamily ligand that triggers the *Drosophila* JNK pathway. EMBO J 21:3009–3018

Joukov V, Sorsa T, Kumar V, Jeltsch M, Claesson-Welsh L, Cao Y, Saksela O, Kalkkinen N, Alitalo K (1997) Proteolytic processing regulates receptor specificity and activity of VEGF-C. EMBO J 16:3898–3911

Khush RS, Leulier F, Lemaitre B (2001) *Drosophila* immunity: two paths to NF-kappaB. Trends Immunol 22:260–264

Lanot R, Zachary D, Holder F, Meister M (2001) Postembryonic hematopoiesis in *Drosophila*. Dev Biol 230:243–257

Lebestky T, Chang T, Hartenstein V, Banerjee U (2000) Specification of *Drosophila* hematopoietic lineage by conserved transcription factors. Science 288:146–149

Li X, Ponten A, Aase K, Karlsson L, Abramsson A, Uutela M, Backstrom G, Hellstrom M, Bostrom H, Li H, Soriano P, Betsholtz C, Heldin C-H, Alitalo K, Ostman A, Eriksson U (2000) PDGF-C is a new protease-activated ligand for the PDGF alpha-receptor. Nat Cell Biol 2:302–309

Locksley RM, Killeen N, Lenardo MJ (2001) The TNF and TNF receptor superfamilies: integrating mammalian biology. Cell 104:487–501

Lotem J, Sachs L (2002) Cytokine control of developmental programs in normal hematopoiesis and leukemia. Oncogene 21:3284–3294

Luo H, Hanratty WP, Dearolf CR (1995) An amino acid substitution in the *Drosophila* hopTum-l JAK kinase causes leukemia-like hematopoietic defects. EMBO J 14:1412–1420
Mackay CR (2001) Chemokines: immunology's high impact factors. Nat Immunol 2:95–101
Meister M, Govind S (2003) Hematopoietic development in *Drosophila*: a parallel with vertebrates. In: Godin I, Cumano A (eds) Hematopoietic stem cell development. Landes Biosc (in press). Eurekahicom
Meister M, Richards G (1996) Ecdysone and insect immunity: the maturation of the inducibility of the diptericin gene in *Drosophila* larvae. Insect Biochem Mol Biol 26:155–160
Melk JP, Govind S (1999) Developmental analysis of *Ganaspis xanthopoda*, a larval parasitoid of *Drosophila melanogaster*. J Exp Biol 202:1885–96
Mizuguchi K, Parker JS, Blundell TL, Gay NJ (1998) Getting knotted: a model for the structure and activation of Spätzle. TIBS 23:239–242
Morey KM, Harri C, Harrison DA (2001) Characterization of the JAK/STAT signaling pathway in *Drosophila*. Annu Drosophila Res Conf 42:418A
Morisato D, Anderson KV (1994) The *spätzle* gene product encodes a component of the extracellular signaling pathway establishing the dorsal ventral pattern of the *Drosophila* embryo. Cell 76:677–688
Mott HR, Campbell ID (1995) Four-helix bundle growth factors and their receptors: protein-protein interactions. Curr Opin Struct Biol 5:114–121
Muller YA, Li B, Christinger HW, Wells JA, Cunningham BC, de Vos AM (1997) Vascular endothelial growth factor: crystal structure and functional mapping of the kinase domain receptor binding site. Proc Natl Acad Sci USA 94:7192–7197
Munier A-I, Daniel Doucet D, Perrodou E, Zachary D, Meister M, Hoffmann JA, Janeway CA, Lagueux M (2002) PVF2, A PDGF/VEGF-like growth factor, induces hemocyte proliferation in *Drosophila* larvae. EMBO J 12:1195–1200
Nicolas E, Nappi AJ, Lemaitre B (1996) Expression of antibacterial peptide genes after infection by parasitoid wasps in *Drosophila*. Dev Comp Immunol 20:175–181
Nicholas KB, Nicholas HB Jr, Deerfield DW II (1997) GeneDoc: analysis and visualization of genetic variation. EMBNEW.NEWS 4:14
Oefner C, D'Arcy A, Winkler FK, Eggimann B, Hosang M (1992) Crystal structure of human platelet-derived growth factor BB. EMBO J 11:3921–3926
Olofsson B, Jeltsch M, Eriksson U, Alitalo K (1999) Current biology of VEGF-B and VEGF-C. Curr Opin Biotech 10:528–535
Parker JS, Mizuguchi K, Gay NJ (2001) A family of proteins related to Spätzle, the Toll receptor ligand, are encoded in the *Drosophila* genome. Proteins 45:71–80
Prestrelski SJ, Arakawa T, Duker K, Kenney WC, Narhi LO (1994) The conformational stability of a non-covalent dimer of a platelet-derived growth factor-B mutant lacking the two cysteines involved in interchain disulfide bonds. Int J Pept Protein Res 44:357–363
Qiu P, Pan PC, Govind S (1998) A role for the *Drosophila* Toll/Cactus pathway in larval hematopoiesis. Development 125:1909–1920
Rane SG, Reddy EP (2002) JAKs, STATs and Src kinases in hematopoiesis. Oncogene 21:3334–3358
Rehorn KP, Thelen H, Michelson AM, Reuter R (1996) A molecular aspect of hematopoiesis and endoderm development common to vertebrates and *Drosophila*. Development 122:4023–4031
Rizki TM (1978) The circulatory system and associated cells and tissues. In: Wright TRF, Ashburner M (eds) Genetics and biology of *Drosophila*, vol 2b. Academic Press, New York, pp 398–451
Rizki RM, Rizki TM (1990) Encapsulation of parasitoid eggs in phenoloxidase-deficient mutants of *Drosophila melanogaster*. J Insect Physiol 36:523–529
Rizki TM, Rizki RM (1992) Lamellocyte differentiation in *Drosophila* larvae parasitized by *Leptopilina*. Dev Comp Immunol 16:103–110
Roth S, Hiromi Y, Godt D, Nüsslein-Volhard C (1991) *cactus*, a maternal gene required for proper formation of the dorsoventral morphogen gradient in *Drosophila* embryos. Development 112: 371–388

Rubin GM, Yandell MD, Wortman JR et al (2000) Comparative genomics of the eukaryotes. Science 287:2204–2215

Rugendorff A, Younossi-Hartenstein A, Hartenstein V (1994) Embryonic origin and differentiation of the *Drosophila* heart. Roux's Arch Dev Biol 203:266–280

Schneider D (2000) Using *Drosophila* as a model insect. Nat Rev Genet 1:218–226

Shrestha R, Gateff E (1982) Ultratructure and cytochemistry of the cell types in the larval hematopoietic organs and hemolymph of *Drosophila melanogaster*. Dev Growth Differ 24: 65–82

Sorrentino RP, Carton Y, Govind S (2002) Cellular immune response to parasite infection in the *Drosophila* lymph gland is developmentally regulated. Dev Biol 243:65–80

Stacker SA, Stenvers K, Caesar C, Vitali A, Domagala T, Nice E, Roufail S, Simpson RJ, Moritz R, Karpanen T, Alitalo K, Achen MG (1999) Biosynthesis of vascular endothelial growth factor-D involves proteolytic processing which generates non-covalent homodimers. J Biol Chem 274:32127–32136

Stein D, Nusslein-Volhard C (1992) Multiple extracellular activities in *Drosophila* egg perivitelline fluid are required for establishment of embryonic dorsal-ventral polarity. Cell 68:429–440

Tauzig S, Jouanguy E, Hoffmann JA, Imler J-L (2000) Toll-related receptors and control of antimicrobial peptide expression in *Drosophila*. Proc Natl Acad Sci USA 97:10520–10525

Tepass U, Fessler LI, Aziz A, Hartenstein V (1994) Embryonic origin of hemocytes and their relationship to cell death in *Drosophila*. Development 120:1829–1837

Thompson JD, Higgins DG, Gibson TJ (1994) CLUSTALW: improving the sensitivity of progressive multiple sequence alignment through sequence weighting, position-specific gap penalties and weight matrix choice. Nucleic Acids Res 22:4673–4680

Ward AC, Touw I, Yoshimura A (2000) The JAK/STAT pathway in normal and perturbed hematopoiesis. Blood 95:19–29

Zon LI (2001) Self-renewal versus differentiation, a job for the mighty morphogens. Nat Immunol 2:142–143

Analogies Between *Drosophila* and Mammalian TRAF Pathways

H. Wajant[1] and P. Scheurich[2]

Abstract. A central event in innate immunity is the activation of the NF-κB signaling pathway and up-regulation of NF-κB-dependent defense genes. Attack of mammals as well as of insects by microorganisms leads, among other things, to the activation of receptors of the Toll-like receptor group. Various adaptor proteins involving members of the TNF receptor-associated factor (TRAF) family channel these receptor-generated signals to conserved intracellular kinase cascades that finally lead to the activation of NF-κB and JNK. In vertebrates, TRAF proteins link these pathways also to IL-1R-related molecules and members of the TNF receptor superfamily, which orchestrate a variety of immunoregulatory processes of the innate but also of the adaptive immune system. In this review, we will focus on the similarities but also the differences in TRAF-dependent signaling pathways of mammals and insects.

1 Introduction

The "TNF receptor-associated factor" (TRAF) family comprises a group of adaptor/scaffold proteins involved in the activation and regulation of the NF-κB pathway or mitogen-activated protein kinase signaling cascades. Currently, six TRAF proteins, designated as TRAF1 to TRAF6, have been described in mammals (Wajant et al. 2001; Chung et al. 2002), three (DTRAF1, DTRAF2 and DTRAF3) in flies (Liu et al. 1999; Grech et al. 2000) and one in *Caenorhabditis elegans* (Wajant et al. 1998) (Fig. 1A). The TRAF proteins are characterized by a carboxy-terminal homology domain, the TRAF domain, which is required for self-association of these molecules. In addition, the TRAF domain enables these proteins to undergo a huge variety of protein – protein interactions. The TRAF domain can be subdivided into the highly conserved TRAF-C subdomain, consisting of an eight-stranded anti-parallel β-sandwich structure and a less conserved amino-terminal part, the TRAF-N domain, which is organized as a coiled-coil (Wajant et al. 2001; Chung et al. 2002). The TRAF domains form trimers reminiscent in their structure to a trefoil, with the three TRAF-C domains as leaves and the trimerized TRAF-N domains as the stalk (Park et al. 1999; McWhirter et al. 1999; Ye et al. 1999; Ni et al. 2000). With the excep-

[1]Department of Molecular Internal Medicine, Medical Polyclinic, University Würzburg, Röntgenring 11, 97070 Würzburg, Germany
[2]Institute of Cell Biology and Immunology, University of Stuttgart, Allmandring 31, 70569 Stuttgart, Germany

Progress in Molecular and Subcellular Biology
A. Beschin (Ed.) Invertebrate Cytokines

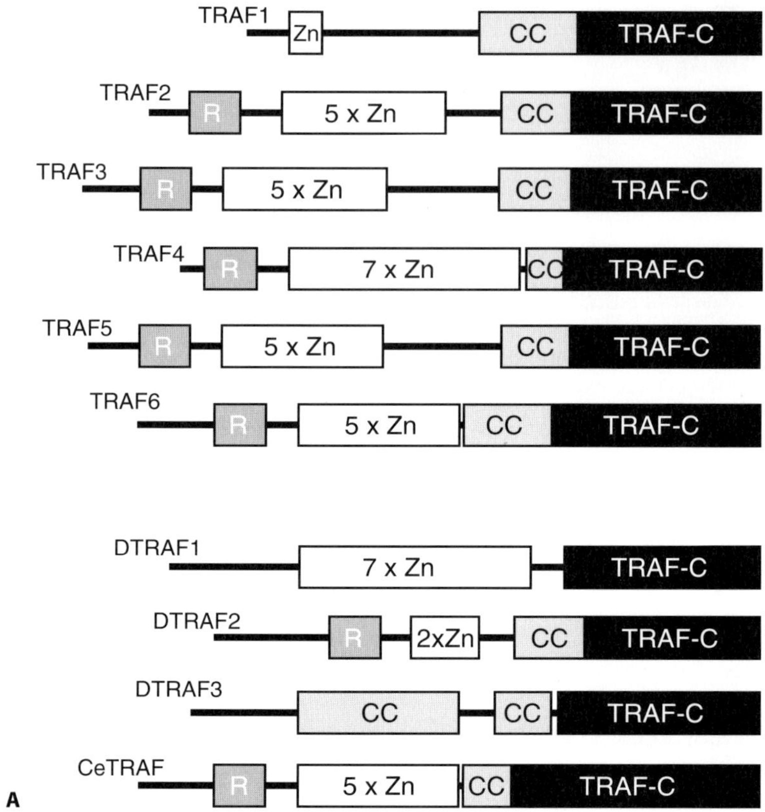

Fig. 1. The TRAF protein family. **A** Domain architecture of TRAF proteins. Note that the carboxy-terminal TRAF-C domain and the foregoing coiled-coil domain (TRAF-N domain) together form the TRAF domain. *R* indicates a RING domain and *Zn* refers to the zinc finger motif. **B** Protein–protein interaction capabilities of a generalized TRAF protein. For details of the plethora of TRAF-interacting proteins, see Wajant et al. (2001)

tion of DTRAF1, DTRAF3 and TRAF1, the amino-terminal parts of all TRAF proteins contain a RING domain followed by a regularly spaced cluster of five or seven zinc fingers (Fig. 1A). While DTRAF1 lacks the RING domain but has also seven zinc fingers, the amino-terminal part of TRAF1 contains a single zinc finger but no additional homologies to the other TRAF proteins. The amino-terminal part of DTRAF3 is also unique among the TRAF proteins and contains an extended coiled-coil region (Grech et al. 2000). As the TRAF domain of TRAF1 is highly homologous to the TRAF domain of TRAF2, the atypical RING/zinc domain-less architecture of the TRAF1 amino-terminal domain most likely arose late in evolution (Zapata et al. 2000; Wajant et al. 2001). There is overwhelming evidence that the TRAF domain of TRAF proteins acts as an assembly platform that interacts with members of the TNF

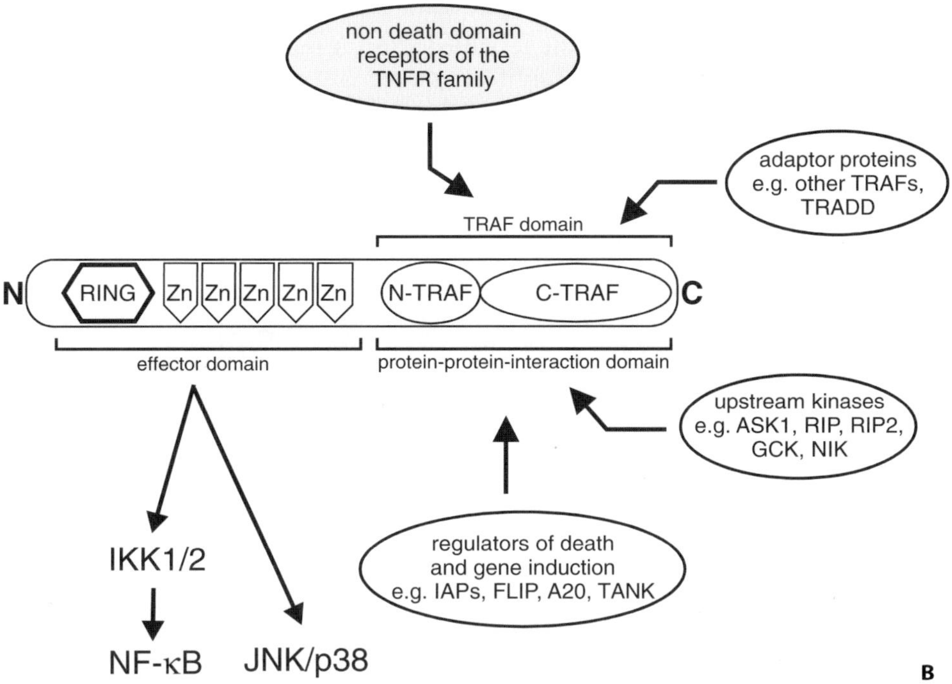

Fig. 1. *Continued*

receptor and IL-1R/Toll receptor families, additional adaptor proteins and a variety of kinases as well as modulatory proteins regulating these kinases (Wajant et al. 2001; Chung et al. 2002; Fig. 1B). In contrast, the role of the amino-terminal RING/zinc domains of TRAF proteins is less understood. It has been recently shown that TRAF6 can act as an E3 ligase most likely catalyzing its own ubiquitination with atypical lysine-63-linked polyubiquitin chains (Deng et al. 2000; Wang et al. 2001). Moreover, ubiquitinated TRAF6 has been implicated in the activation of TGF-β-activated kinase (Tak1), a mitogen-activated protein kinase kinase kinase (MAP3K) involved in the stimulation of the NF-κB-inducing I-κB kinase (IKK) complex (Wang et al. 2001). In addition, for TRAF2 and TRAF6, interactions of their RING/zinc domains with MEKK1 (Baud et al. 1999), another MAP3K, and IKK2 (Devin et al. 2001), a subunit of the IKK complex, have been found. Overexpression of TRAF proteins leads to the activation of TRAF-associated signaling pathways. It is noteworthy that, while the expression of the sole RING/zinc domains is not sufficient to trigger the activation of TRAF-associated signaling pathways, the artificial oligomerization of these domains restores signaling capacities attributed to the corresponding complete TRAF proteins (Baud et al. 1999; Wang et al. 2001). Taken together, these findings suggest that TRAF proteins not only act as scaffolds to bring together receptors and cytosolic signaling intermediates, but may also have a role in the activation of the recruited effector molecules. In recent years, additional proteins have been identified that contain domains with significant homologies to the TRAF domain of the "classical" TRAF proteins, the so-called TEFs (TRAF domain-encompassing factors) (Zapata et al. 2001). However, the role of these TEFs and their relationship to TRAF-associated functions are still completely unclear.

2 TRAF-Dependent Signaling Pathways in Mammalian and Insect Innate Immunity

There is a common need of vertebrates and invertebrates to protect themselves against the attack of a wide variety of pathogens. Both mammals and insects have developed a range of mechanisms enabling them to respond rapidly against potentially harmful microorganisms. This innate immunity response is essentially based on the sensing of microorganism-derived substances by germ-line encoded pattern recognition proteins and induction of defense mechanisms, in particular activation of phagocytic cells and transcriptional up-regulation of anti-microbial peptides. In insects, anti-microbial peptides are either induced in epithelial cells to act locally or are produced in the fat body, the insect analogue of the vertebrate liver, to act systemically (Meister et al. 2000). In mammals, anti-microbial peptides are predominantly expressed in epithelial tissue and comprise constitutively expressed as well as inducible forms (Yang et al. 2001; Lehrer and Ganz 2002). In vertebrates, innate immunity is additionally tightly linked to the adaptive immune system that responds to pathogens with the selective expansion of antigen-specific T and B cells, which are selected from a huge variety of clones generated by somatic chromosomal DNA rearrangement in the T- and B-cell receptor genes. In *Drosophila*, seven anti-microbial peptides exist: drosomycin, which is predominantly active against fungi, metchnikowin, which exerts anti-fungal and anti-bacterial activity, and attacin, cecropin, defensin, diptericin and drosocin, which mainly act against bacteria (Meister et al. 2000). Infection-induced transcriptional up-regulation of these anti-microbial peptides is achieved by two distinct signaling pathways: the immune deficiency (Imd)-related pathway (Lemaitre et al. 1995) and the Toll pathway (Lemaitre et al. 1996). Both pathways lead to the activation of members of the NF-κB/Rel transcription factor family (see Appendix 1) but are well separated upstream of these factors. Thus, mutations affecting/inhibiting a component of one pathway do not interfere with the functionality of the other pathway (Table 1). While the Toll pathway predominantly induces drosomycin and the Imd pathway up-regulates selectively diptericin and drosocin, the remaining anti-microbial peptides require both pathways for maximal induction (Lemaitre et al. 1996; Levashina et al.

1998). Distinct categories of microbes differentially stimulate the Toll and Imd pathways. The Toll pathway is triggered by fungi or Gram-positive bacteria, whereas the Imd pathway is activated by Gram-negative bacteria (Silverman and Maniatis 2001; Hoffmann and Reichhart 2002). Microbial infection also leads to activation of the c-Jun-N-terminal kinase (JNK) (see Appendix 2) and JAK/STAT pathways (Sluss et al. 1996; Lagueux et al. 2000; Levashina et al. 2001). While the latter induces up-regulation of complement-related proteins implicated in the opsonization of Gram-negative bacteria, the importance of the JNK pathway for the *Drosophila* innate immune response is still unclear.

With regard to intracellular signaling intermediates, the Imd and Toll pathways in *Drosophila* are non-redundant despite utilizing in principally related pathways. Remarkably, phylogenetically conserved components of both the Imd and the Toll pathways work in combination in the mammalian Toll-like receptors/IL-1R pathway (see below). Noteworthy, the Imd pathway utilizes in addition several conserved proteins, which have been implicated in apoptosis-induction in mammals and *Drosophila*.

2.1 The Toll Pathway in Innate Immunity of *Drosophila*

The Toll pathway has been originally identified due to its pivotal role in the dorsoventral patterning of the *Drosophila* embryo (Belvin and Anderson 1996). Toll is a transmembrane receptor and the name-giving part of a signaling pathway that transduces extracellular "information" to the nucleus to initiate a transcriptional response towards these extracellular stimuli (Hashimoto et al. 1988). Toll consists of an extracellular domain comprising several leucine-rich repeats, a single transmembrane domain and an intracellular part containing a conserved domain of about 150 amino acids. The conserved intracellular domain is also found in other Toll-like receptors (TLRs), in the IL-1R family as well as in the MyD88 adaptor proteins and is therefore named the Toll/IL-1R homology (TIR) domain (Silverman and Maniatis 2001). In dorsoventral patterning of the *Drosophila* embryo, Toll gets activated by a cleavage product of the cysteine knot growth factor-like protein Spätzle, which consists of a disulfide-linked dimer of the carboxy-terminal 106 amino acids of the complete Spätzle protein (Morisato and Anderson 1994; DeLotto and DeLotto 1998). Proteolytic activation of Spätzle in dorsoventral patterning is achieved by a serine protease cascade comprising Gastrulation Defective, Snake, and Easter (LeMosy et al. 1999). Proteolytic activation of Spätzle has also an essential role in activation of Toll in innate immunity, but in this case this process is independent from the aforementioned protease cascade (Lemaitre et al. 1996; Levashina et al. 1999). Thus, loss-of-function mutants of Gastrulation Defective, Snake, and Easter have no inhibitory effect on Toll-mediated up-regulation of drosomycin during infection with Gram-positive bacteria or fungi, whereas a loss-of-function mutant of Spätzle strongly reduces these

Table 1. Innate immunity in mutant *Drosophila* strains

Gene	Type of mutation	Inducible activity		Constitutive activity		References
		Toll	Imd	Toll	Imd	
Wild type		Yes	Yes	No	No	
semmelweis (PGRP-SA)	Loss-of-function	No (Gram-positive) Yes (fungi)	Yes	No	No	Michel et al. (2001)
semmelweis (PGRP-SA)	Loss-of-function in toll gain-of-function flies	n.d.	n.d.	Yes	No	Michel et al. (2001)
semmelweis (PGRP-SA)	Loss-of-function in necrotic loss-of-Function flies	n.d.	n.d.	Yes	No	Michel et al. (2001)
necrotic (Spn43ac)	Loss-of-function	Yes	Yes	Yes	No	Levashina et al. (1999)
necrotic (Spn43ac)	Loss-of-function in toll or spätzle or snake or gastrulation defective loss-of-function flies	No	Yes	No	No	Levashina et al. (1999)
easter	Loss-of-function	Yes	Yes	No	No	Lemaitre et al. (1996)
snake	Loss-of-function	Yes	Yes	No	No	Lemaitre et al. (1996)
spätzle	Loss-of-function	No	Yes	No	No	Lemaitre et al. (1996)
toll	Loss-of-function	No	Yes	No	No	Lemaitre et al. (1996)
toll	Gain-of-function	Yes	Yes	Yes	No	Lemaitre et al. (1996)
dmyd88	Overexpression in wild-type flies			Yes	No	Tauszig-Delamasure et al. (2002)
dmyd88	Overexpression in pelle loss-of-function flies			No	No	Tauszig-Delamasure et al. (2002)
dmyd88	Overexpression in tube loss-of-function flies			No	No	Tauszig-Delamasure et al. (2002)
pelle	Loss-of-function	No	Yes	No	No	Lemaitre et al. (1996)
tube	Loss-of-function	No	Yes	No	No	Lemaitre et al. (1996)
cactus	Loss-of-function	Yes	Yes	Yes	No	Lemaitre et al. (1996)
dif	Loss-of-function	Strongly reduced	Yes	No	No	Rutschmann et al. (2000a)
dif	Loss-of-function in cactus loss-of-function flies	n.d.	n.d.	No	n.d.	Rutschmann et al. (2000a)

ird7 (PGRP-LC)	Loss-of-function	Normal	Strongly reduced	No	No	Choe et al. (2002); Ramet et al. (2002)
imd	Overexpression in imd loss-of-function flies	Yes	Yes	No	Yes	Georgel et al. (2001)
imd	Overexpression in kenny loss-of-function	Yes	No	No	No	Georgel et al. (2001)
imd	Overexpression in dredd loss-of-function	Yes	No	No	No	Georgel et al. (2001)
dtak1	Loss-of-function	Yes	No	No	No	Vidal et al. (2001)
dtak1	Heat shock-induced in wild-type flies			No	Yes	Vidal et al. (2001)
dtak1	Heat shock-induced in ird5 Loss-of-function flies			No	No	Vidal et al. (2001)
dtak1	Heat shock-induced in kenny loss-of-function flies			No	No	Vidal et al. (2001)
dtak1	Heat shock-induced in imd loss-of-function flies			No	Yes	Vidal et al. (2001)
kenny	Loss-of-function	Yes	No	No	No	Rutschmann et al. (2000a)
ird5	Loss-of-function	Yes	No	No	No	Lu et al. (2001)
dredd	Loss-of-function	Yes	No	No	No	Leulier et al. (2000)
dredd	Heat shock-induced in wild type flies			No	Yes	Vidal et al. (2001)
dredd	Heat shock-induced in kenny loss-of-function flies			No	Yes	Vidal et al. (2001)
dredd	Heat shock-induced in imd loss-of-function flies			No	Yes	Vidal et al. (2001)
relish	Loss-of-function	Yes	No	No	No	Hedengren et al. (1999)
toll + imd	Both loss-of-function	No	No	No	No	Lemaitre et al. (1996); Levashina et al. (1998)

responses (Lemaitre et al. 1996). In agreement with a distinct proteolytic cascade controlling Spätzle activation in Toll-dependent signaling in innate immunity, Spn43Ac, a serine protease inhibitor of the serpin family, prevents constitutive processing of Spätzle, but has no effect on dorsoventral patterning or the Imd pathway (Levashina et al. 1999). Recognition of Gram-positive bacteria and fungi and activation of Toll diverges upstream of Spätzle processing. This became evident from the identification of a soluble peptidoglycan recognition protein (PGRP-SA) that is essentially involved in Toll activation by several Gram-positive bacteria, but has no role in the anti-fungal response (Michel et al. 2001). How recognition of Gram-positive bacteria by PGRP-SA results in the processing of Spätzle remains to be elucidated. The Spätzle processing-dependent mode of Toll activation in innate immunity of *Drosophila* suggests that Toll itself does not directly recognize microbe-derived molecular patterns. This is in striking contrast to innate immunity in mammals where free or protein complexes of microbial products directly stimulate TLRs without the need of activated proteases (Akira et al. 2001). Moreover, in mammals, all TLRs seem to be involved in innate immunity, whereas in *Drosophila*, there is yet no evidence for an anti-microbial function of most of the Toll-related receptors (Tauszig et al. 2000).

Activation of Toll results in the recruitment of the *Drosophila* homologue of the mammalian adapter protein "myeloid differentiation primary response gene 88" (DmMyD88), the serine/threonine kinase Pelle and Tube. Loss-of-function mutants of these proteins interfere with induction of the Toll pathway-specific peptide drosomycin upon fungal or Gram-positive bacteria infection but have no impact on the Imd pathway-mediated response (Lemaitre et al. 1996; Tauszig-Delamasure et al. 2002). Like its mammalian counterpart, DmMyD88 consists of an amino-terminal death domain followed by a TIR domain and a phylogenetically non-conserved carboxy-terminal extension. DmMyD88 binds Toll by homophilic interaction of the TIR domains of both proteins (Horng and Medzhitov 2001; Tauszig-Delamasure et al. 2002). DmMyd88 is also able to bind Pelle (Horng and Medzhitov 2001; Tauszig-Delamasure et al. 2002), which itself interacts directly with Toll in its non-phosphorylated form by its carboxy-terminal catalytic domain (Shen and Manley 1998). In addition, Pelle interacts with the amino-terminal death domain of Tube by its own amino-terminal death domain (Grosshans et al. 1994; Galindo et al. 1995). Again, this interaction is predominantly observed with non-phosphorylated Pelle (Edwards et al. 1997). Pelle phosphorylates Toll, Tube and itself in vitro (Grosshans et al. 1994; Shen and Manley 1998). Expression of a kinase-inactive mutant of Pelle results in increased localization of Tube to sites of Toll receptor signaling in *Drosophila* embryos, suggesting that Pelle-mediated phosphorylation events trigger the release of one or more components from the Toll signaling complex (Towb et al. 2001). Up-regulation of drosomycin by overexpression of DmMyD88 is abrogated in flies with loss-of-function mutants of Tube or Pelle, placing DmMyD88 upstream of these molecules in the Toll pathway (Tauszig-Delamasure et al. 2002; Fig. 2).

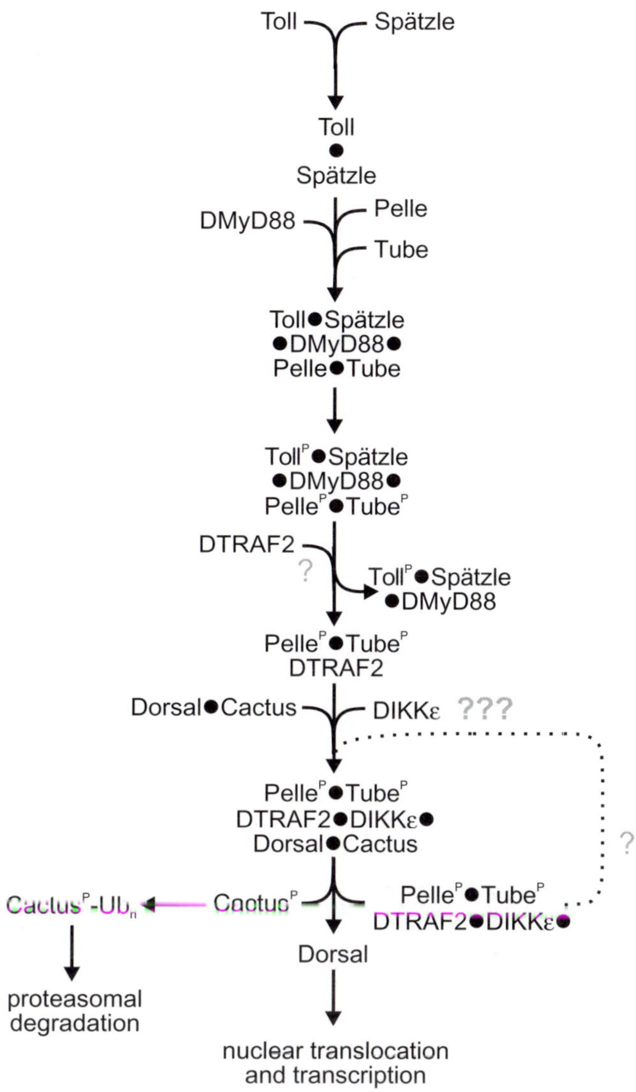

Fig. 2. Model of Spätzle-mediated activation of Dorsal. Events in the pathway currently not directly substantiated by experimental evidence are labeled with "?"

Signal initiation by activated IL-1R or mammalian TLRs is very similar to the mechanisms described above for activated Toll. Again, the adapter protein MyD88 is recruited to the activated receptor by homophilic TIR domain interaction (Muzio et al. 1997; Wesche et al. 1997; Burns et al. 1998; Medzhitov et al. 1998). Analysis of MyD88 knockout mice revealed an essential role of this adaptor protein for signal transduction by the receptors of IL-1 and IL-18 as well as for some of the TLRs (Adachi et al. 1998; Kawai et al. 1999; Takeuchi et al. 2000). Moreover, there is evidence that the MyD88-independent TLRs utilize the MyD88-related protein MAL (MyD88-adapter-like) in a similar way (Fitzgerald et al. 2001). In addition, the functions of Pelle and Tube in Toll signaling are at least partly mirrored by the mammalian homologues of Pelle, the members of the "IL-1 receptor-associated kinase" (IRAK) family. In humans, four IRAKs have been identified: IRAK-1 (Cao et al. 1996), IRAK-2 (Muzio et al. 1997), IRAK-M (Wesche et al. 1999) and IRAK-4 (Li et al. 2002). The IRAKs are recruited to the IL-1R in a ligand-dependent manner and, like Pelle, interact with the concomitantly recruited MyD88 protein (Cao et al. 1996; Muzio et al. 1997; Wesche et al. 1999; Li et al. 2002; Fig. 3). Reconstitution experiments in IRAK-1-deficient cells, where IL-1-induced activation of NF-κB and JNK is strongly reduced, revealed that IRAK-M and IRAK-2 are able to replace IRAK-1 in IL-1 signaling (Li et al. 1999; Wesche et al. 1999). These studies further revealed that the kinase activity of these enzymes is dispensable for their role in IL-1 signaling. Nevertheless, kinase-dead mutants of IRAK-1, IRAK-2 or IRAK-M are phosphorylated and this is tightly correlated with the capability of these mutants to induce NF-κB (Li et al. 1999). Like its *Drosophila* counterpart, MyD88 acts downstream of the receptor but upstream of IRAK-4 and IRAK-1 (Suzuki et al. 2002). It has been recently shown that IRAK-4 phosphorylates IRAK-1 in response to IL-1 (Li et al. 2002). Moreover, the kinase activity of IRAK-4 is indispensable for its NF-κB-inducing capacity (Li et al. 2002). In agreement with the functional redundancy of IRAK-1, IRAK-2 and IRAK-M and the unique role of IRAK-4 in IL-1-mediated NF-κB activation, mice deficient in IRAK-1 show residual IL-1-induced NF-κB activity (Kanakaraj et al. 1998) whereas mice lacking IRAK-4 are completely blocked in this response (Suzuki et al. 2002). So far investigated, IRAK-1 and IRAK-4 have a similar function in the signal transduction by some of the human TLRs and the IL-18R (Kanakaraj et al. 1998; Suzuki et al. 2002). Taken together, there is a high degree of similarity in the molecular mode of function between MyD88 and IRAK-4 in mammalian cells and DmMyD88 and Pelle in *Drosophila*. Moreover, with respect to their interplay with MyD88/DmMyD88 and IRAK-4/Pelle, the phylogenetically unrelated proteins IRAK-1 and Tube also act in a comparable manner. Phosphorylation of IRAK-1 in the receptor complex leads to the interaction with the adaptor protein TRAF6 and the dissociation of both molecules from the IL-1R complex (Qian et al. 2001). Remarkably, there is evidence that Pelle is involved in a similar way in the release of Tube from the Toll complex (Towb et al. 2001). Phosphorylation of IRAK-1 may also result in its ubiquitination and proteasomal degradation (Yamin and Miller 1997), but these events are apparently dispensable for its NF-κB-inducing function (Qian et al. 2001).

Three members of the TRAF protein family have been recently identified in *Drosophila* (DTRAF1, DTRAF2, DTRAF3), but the potential involvement of these proteins in the Toll pathway is poorly understood. In agreement with a role of DTRAF2 in Toll signaling, overexpression of this TRAF protein was suf-

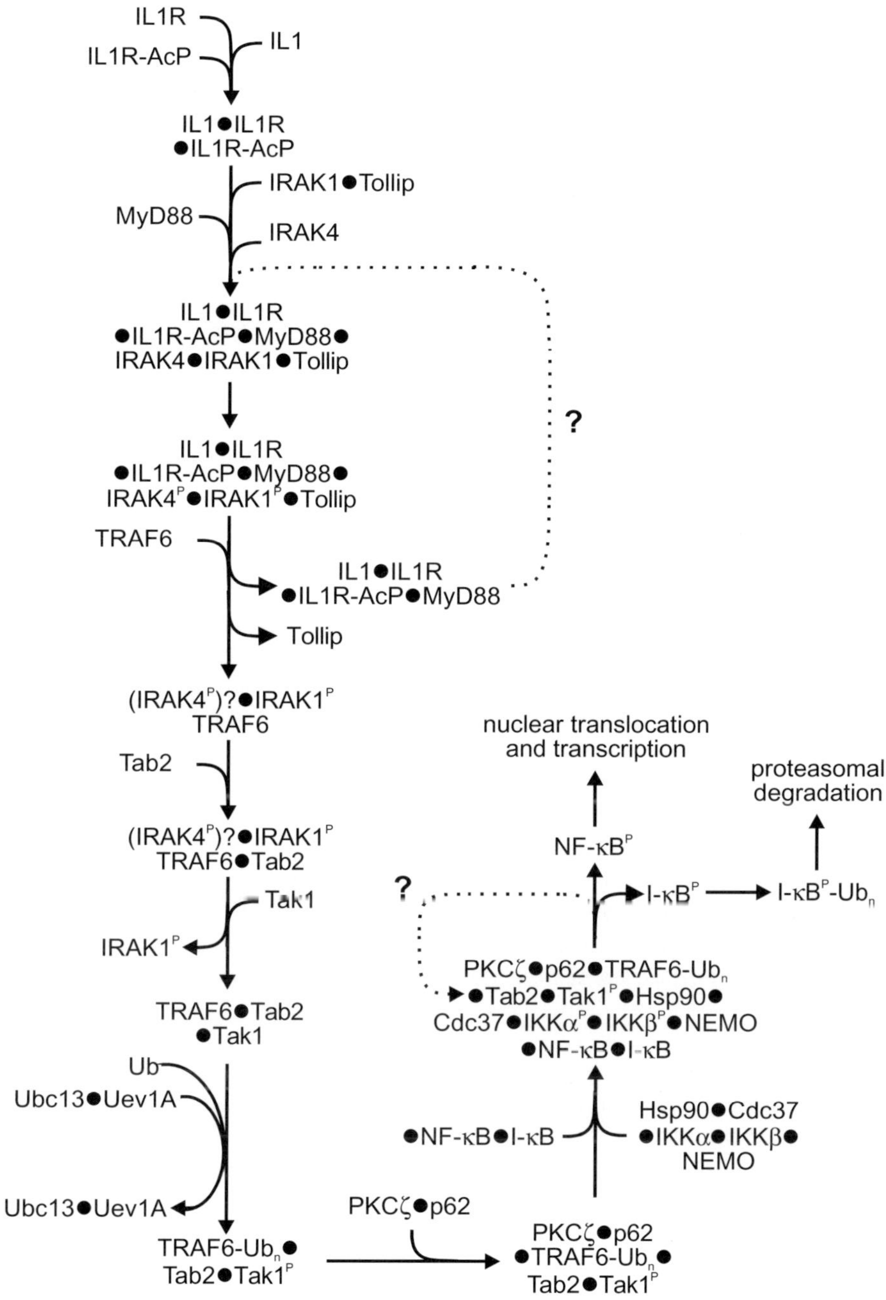

Fig. 3. Model of IL-1-induced activation of NF-κB. Events in the pathway currently not directly substantiated by experimental evidence are labeled with "?"

ficient to drive a Dorsal-regulated reporter gene (Shen et al. 2001). Moreover, DTRAF2 and Pelle synergistically activate Dorsal and a weak interaction of both molecules can be observed in vitro (Shen et al. 2001). Deletion mutant analyses of Pelle showed that it interacts with DTRAF2 via its kinase domain (Shen et al. 2001). Interestingly, the RING/zinc finger domain of DTRAF2 alone interacts much stronger with Pelle as compared to the complete protein (Shen et al. 2001). Moreover, the RING/zinc domain of DTRAF2 is more potent than complete DTRAF2 in the activation of a Dorsal-driven reporter gene, and synergizes, in this regard, with Pelle too. An amino-terminal truncated deletion mutant of DTRAF2, solely comprising the TRAF domain of the protein, is nevertheless able to activate Dorsal (Shen et al. 2001). This is very surprising because in mammalian cells, the corresponding TRAF6 deletion mutant, but also similar mutants of other TRAF proteins, displayed a dominant-negative effect on the respective signaling pathways. Thus, the interaction of TRAF6 and IRAK-1 or IRAK-2 relies on the TRAF domain of TRAF6 (Muzio et al. 1997) and the amino-terminal regulatory domain of IRAK-1 and IRAK-2 (Li et al. 2001), whereas the interaction of DTRAF2 and Pelle seems to be based on the RING/zinc finger domain of DTRAF2 and the kinase domain of Pelle (Shen et al. 2001). Therefore, at a first glance, the molecular mode of TRAF – kinase interaction seems not to be conserved between mammals and flies. However, for the reasons outlined above, IRAK-4 may be the only kinase of the IRAK family that fulfills a Pelle analogue role in mammals, whereas the other IRAK family members functionally behave like *Drosophila* Tube. Thus, it remains to be clarified (1) whether there is an interaction between Tube and DTRAF2, qualitatively reflecting the IRAK-1 interaction with TRAF6, and (2) whether there is an interaction between the kinase domain of IRAK-4 and the RING/zinc finger domain of TRAF6.

Although mammalian TRAF4 seems not to be involved in NF-κB signaling, its *Drosophila* counterpart DTRAF1 has been implicated in Toll signaling as it interacts via its TRAF domain with the death domain-containing regulatory domain of Pelle (Zapata et al. 2000). Thus, DTRAF1 and DTRAF2 utilize different domains (TRAF domain versus RING/zinc finger domain) to interact with different parts of Pelle (regulatory domain versus kinase domain). Although an amino-terminal deletion mutant of DTRAF1 interferes with Pelle-induced NF-κB activation upon overexpression in mammalian cells, microinjection of the respective mRNA in *Drosophila* had no effect on dorsoventral patterning (Zapata et al. 2000), arguing against a major role of DTRAF1 in Toll-induced activation of NF-κB/Dorsal. DTRAF1 was originally identified due to its interaction with Misshapen, a mitogen-activated protein kinase kinase kinase kinase (MAP4K) that is involved in *Drosophila* in the activation of the JNK mitogen-activated protein kinase module (Liu et al. 1999). It is therefore tempting to speculate that DTRAF1 is involved in Toll-mediated activation of the JNK pathway. Indeed, ectopic expression of Tube and Pelle activates the JNK pathway in mammalian cells (Bacher et al. 2001). Moreover, it has been shown that the *Drosophila* homologue of JNK can be activated by LPS, pointing to a role of JNK in the Imd pathway too (Sluss et al. 1996). Similarly, bifurcation of the *Drosophila* JNK and NF-κB signaling pathways at the level of Pelle may also occur in mammalian TLR signaling as a dominant-negative mutant of TRAF6 interferes with NF-κB activation by TLRs, but failed to block activation of JNK (Muzio et al. 1998). However, at least in IL-1R and IL-17R signaling the situation seems to be different, because in TRAF6-deficient mice, both NF-κB and JNK activations are impaired in response to IL-1 and IL-17 (Lomaga et al. 1999; Schwandner et al. 2000). The TRAF6 knockout mice have also a defect in LPS/TLR-induced NF-κB activation, but unfortunately TLR-induced JNK activation was not investigated in these mice yet. Nevertheless, even in the case of IL-1R and IL-17R signaling, IRAK-1 seems to be the bifurcation point of NF-κB and JNK activation as different domains of the molecule are required to mediate these responses (Li et al. 2001).

In the IL-1-induced NF-κB pathway, TRAF6 most likely has multiple, yet poorly understood functions. TRAF6 acts as an E3 ligase which catalyses, together with a dimeric ubiquitin-conjugating enzyme complex consisting of Ubc13 and Uev1A, its own ubiquitination (Deng et al. 2000; Wang et al. 2001). Ubiquitinated TRAF6 then facilitates Tak1-dependent activation of the IKK complex (Wang et al. 2001). Remarkably, polyubiquitin chains are unusually linked to TRAF6 through lysine-63 (K63), instead of lysine-48 (K48) (Deng et al. 2000; Wang et al. 2001). While ubiquitination of TRAF6 is necessary for Tak1-dependent activation of the IKK complex, proteasomal degradation, which normally occurs with K48 – ubiquitin linked proteins, is dispensable (Deng et al. 2000). Noteworthy, IFN-γ induces rapid proteasomal degradation of TRAF6 in osteoclasts, opening the possibility that TRAF6 is also the target of K48 ubiquitination (Takayanagi et al. 2000). Nevertheless, how Tak1 is activated by ubiquitinated TRAF6 is still unknown. DTRAF2, and deletion mutants of DTRAF2 containing the RING domain, are modified upon transient expression in *Drosophila* SL2 cells (Shen et al. 2001). As this modification has been ruled out to be phosphorylation but leads to a ladder-like staining in western blotting experiments (Shen et al. 2001), most likely DTRAF2 is also ubiquitinated during Toll signaling. However, a role of DTRAF2 modification/ubiquitination for Dorsal activation or DTRAF2 degradation remains to be established. TRAF6 also acts, together with p62, as an adapter joining the IKK complex and the atypical protein kinase C zeta (PKCζ), a kinase involved in transactivation of p65 and/or tissue-specific activation of IKKs (Sanz et al. 2000; Leitges et al. 2001). A similar p62-like adaptor function has also been proposed for the phylogenetically conserved ECSIT (evolutionary conserved signaling intermediate in Toll pathway) protein, which concomitantly interacts with MEKK1 and TRAF6/DTRAF2 (Kopp et al. 1999). In *Drosophila*, a homologue of Tak1 (DTak1) exists, but surprisingly, the DTak1 protein is essentially involved in the Imd pathway and dispensable for Toll-induced activation of NF-κB (Vidal et al. 2001). In IL-1 signaling, TRAF6, which is predominantly localized at the plasma membrane (Qian et al. 2001), is not the only protein that translocates into the cytosol in complex with phosphorylated IRAK-1. Likewise, Tab2, a primarily membrane-associated protein acting as an adaptor for TRAF6 interaction with Tab1 and Tak1, behaves in a similar way (Takaesu et al. 2001). Thus, IL-1-induced phosphorylation of IRAK-1 results in the appearance of a cytosolic complex containing phosphorylated IRAK-1, TRAF6, Tak1 and Tab2, the latter being involved in the activation of Tak1. It is currently unknown whether DTak1 is dependent in its function on Tab1/2-related factors.

TRAF6-mediated activation of Tak1 leads to the activation of the kinases of the IKK complex. The activated IKK complex phosphorylates I-κB on conserved serine residues thereby marking this protein for ubiquitination and subsequent proteasomal degradation (see Appendix 1). In non-stimulated cells, I-κB is complexed with dimers of the Rel/NF-κB family of transcription factors (see Appendix 1). Thus, as a consequence of I-κB degradation, NF-κB transcription factors are released to translocate to the nucleus. Likewise, activation of the Toll pathway in *Drosophila*, results in the phosphorylation of the regulatory domain of the *Drosophila* I-κB homologue Cactus and its subsequent proteasomal degradation (Spencer et al. 1999). The microbial activation of the Toll pathway in larvae causes Cactus degradation, and results in nuclear translocation of Dorsal and Dif, two Rel/NF-κB proteins, whereas in adult flies only Dif is required for the anti-fungal response (Manfruelli et al. 1999; Meng et al. 1999; Rutschmann et al. 2000a). The Cactus kinase, which would close the gap between Pelle-DTRAF2 interaction and Cactus phosphorylation, has not been identified yet. DLAK/DmIKKβ (Kim et al. 2000; Silverman et al. 2000) and Kenny/DmIKKγ (Rutschmann et al. 2000b; Silverman et al. 2000) have been recently described in *Drosophila* as homologues of mammalian IKKβ and NEMO/IKKγ, both essentially involved in IL-1R/Toll receptor family-induced NF-κB activation (Karin and Lin 2002). Surprisingly, both molecules have been found to be dispensable for the Toll immune response but have an essential role in the Imd pathway (Table 1), which mediates proteolytical activation of the Rel/NF-κB protein Relish in response to infection with Gram-negative bacteria. However in mammals, a PMA-responsive alternative IKK complex lacking IKKα, IKKβ and NEMO/IKKγ has been implicated in T-cell activation, demonstrating the existence of at least two independent IKK complexes (Peters et al. 2000). This novel IKK complex contains a yet unidentified kinase phosphorylating serine-32 (S32) and serine-36 (S36) of I-κB, and the LPS-inducible IKK-related kinase IKKε/IKKi phosphorylating only S36 of I-κB (Shimada et al. 1999; Peters et al. 2000). There is an additional IKKε/IKKi-like kinase designated as NAK/TBK1/T2 K in mammals that selectively phosphorylates S36 of I-κB and seems to have a role in activation of the "classical" IKKα/IKKβ/NEMO complex (Pomerantz and Baltimore 1999; Bonnard et al. 2000; Tojima et al. 2000). The *Drosophila* DNA database revealed in addition to DmIKKβ only one IKK-like kinase, with closest homologies to IKKε/IKKi and TBK/NAK/T2 K. Thus, it is likely that the *Drosophila* homologue of IKK-related kinases may be part of a Cactus kinase complex involved in Toll signaling acting independently from the DmIKKβ/DmIKKγ complex.

2.2 The Imd Pathway in Innate Immunity of *Drosophila*

The *Drosophila* Imd pathway, which is activated in response to Gram-positive bacteria, leads to the activation of Relish, a member of the Rel/NF-κB family (Fig. 4A). In addition to its amino-terminal Rel-homology domain, Relish has six carboxy-terminal ankyrin repeats holding the molecule inactive in the cytoplasm of non-induced cells (Dushay et al. 1996; see Appendix 1). Thus, the inhibitory carboxy-terminal domain of Relish fulfills a similar function as Cactus in the Toll pathway. Activation of Relish relies on the inducible endoproteolytical cleavage of the molecule, releasing the amino-terminal RHD from the inhibitory ankyrin repeats, thus allowing the translocation of the former into the nucleus (Silverman et al. 2000; Stoven et al. 2000). In mammals, the p52 NF-κB subunit is released in a strikingly similar signal-induced manner

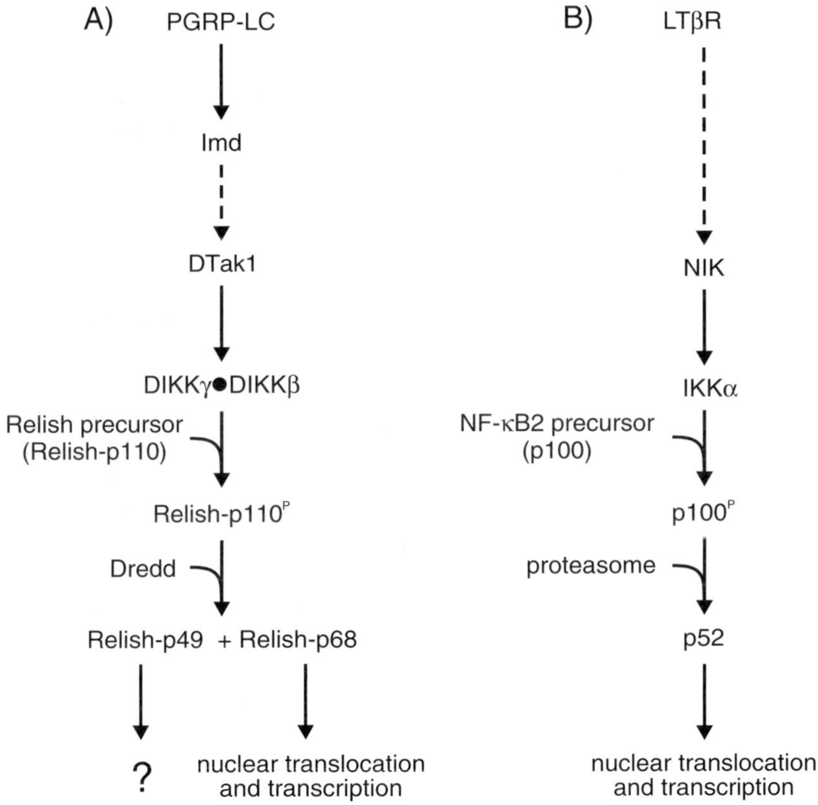

Fig. 4. Model of signal-induced processing of Rel/NF-κB precursor proteins **A** *Drosophila* and **B** mammalian

from its p100 precursor (Fig. 4B). The link in the Imd pathway between Gram-negative bacteria in the extracellular space and the intracellular signaling pathway leading to Relish processing is based on the peptidoglycan recognition protein PGRP-LC (Gottar et al. 2002; Ramet et al. 2002). PGRP-SA, which triggers the Toll pathway upon infection with Gram-positive bacteria, and PGRP-LC belong to a large family of phylogenetically conserved proteins involved in the recognition of peptidoglycans (Werner et al. 2000; Liu et al. 2001). In contrast to PGRP-SA which is a soluble protein, PGRP-LC is a type II transmembrane protein with three peptidoglycan-binding domains in its extracellular part and a large intracellular domain lacking homology to proteins of known function (Gottar et al. 2002; Ramet et al. 2002). The name-giving Imd protein acts downstream of PGRP-LC and has recently been identified as a cytosolic protein with a carboxy-terminal death domain (Appendix 3) showing significant homology to the death domains of the mammalian adaptor proteins TRADD, FADD and RIP, the latter also exerting serine/threonine kinase activity (Georgel et al. 2001). These mammalian adaptor proteins link death domain-containing cell surface receptors to intracellular signaling cascades and lead to apoptosis-induction and/or NF-κB activation. It is therefore tempting to speculate that Imd also acts as a receptor-proximal adaptor coupling PGRP-LC to Relish activation. Noteworthy, both molecules, Imd and its closest mammalian homologue RIP, have been implicated in activation of Rel/NF-κB members, but also in the induction of cell death (Bodmer et al. 2000; Georgel et al. 2001). The next molecular event in Imd-mediated activation of Relish is activation of the MAP3K DTak1 (Vidal et al. 2001). DTak1 then in turn stimulates the *Drosophila* IKK complex, which phosphorylates Relish and thereby induces its endoproteolytic activation (Rutschmann et al. 2000b; Silverman et al. 2000). In a similar way, the mammalian MAP3 K NF-κB-inducing kinase (NIK) stimulates the IKK complex, which in turn activates the NF-κB2 precursor p100 (Senftleben et al. 2001; Xiao et al. 2001a,b) (Fig. 4B). However, with respect to the proteolytic activities involved in the processing of p100 and Relish, there is a striking difference: p100 is processed via an ubiquitination-dependent process by the proteasome (Heusch et al. 1999) whereas Relish is cleaved by Dredd (Leulier et al. 2000), a caspase otherwise involved in apoptosis induction (Chen et al. 1998). How the IKK complex-activating triple kinases DTak1 and NIK are linked to receptors triggering processing of Relish and p100 is still unclear. Again one possibility is that adaptor proteins of the TRAF family are involved. Indeed, in mammals, both NIK and Tak1 can form a complex with TRAF proteins (Song et al. 1997; Ninomiya-Tsuji et al. 1999). Moreover, the Imd-related mammalian adaptor proteins TRADD and RIP interact with TRAF proteins and members of the TNF receptor family, thereby linking these receptors in a TRAF-dependent manner to the activation of the NF-κB pathway (Wajant and Scheurich 2001). However, an involvement of TRAF proteins in receptor-induced processing of Relish or p100 has not been shown yet and future studies must reveal whether such a connection indeed exists.

3
Toll and TRAF-Related Signaling in *Caenorhabditis elegans*

Although putative members of the Toll/IL-1R, Pelle/IRAK, Cactus/I-κB and TRAF protein family have been identified in *C. elegans* (Wajant et al. 1998; Rich et al. 2000; Pujol et al. 2001), other components involved in Toll-induced NF-κB signaling, like the Rel/NF-κB proteins itself or components of the IKK complex, appear to be absent in this species (Pujol et al. 2001). It is noteworthy that none of the aforementioned homologues has been implicated in *C. elegans* resistance against microbial challenge (Pujol et al. 2001). However, a null allele of *C. elegans* Toll (CeToll) has a developmental phenotype whereas putative null alleles of homologues of Pelle/IRAK, Cactus/I-κB and TRAF exert no obvious phenotypes (Pujol et al. 2001). Thus, the molecular basis of developmental signaling by CeToll and *Drosophila* Toll are clearly distinct. In agreement with this difference, an allele of CeToll with a defect in the conserved TIR domain, which in vertebrates and arthropods is involved in coupling Toll receptors to the NF-κB pathway, has only a minor developmental defect (Pujol et al. 2001). Remarkably, worms with the TIR domain-impaired Toll allele have lost their capability to distinguish between different bacterial species and are unable to avoid certain pathogenic bacteria. Again, the *C. elegans* homologues of Toll/IL-1R, Pelle/IRAK, Cactus/I-κB and TRAF are apparently not involved in this process. It is tempting to speculate that CeToll, like some other members of this receptor family, is involved in recognition of specific bacterial patterns. An obvious question is whether *C. elegans* has secondary lost parts of a formerly existing Rel/NF-κB pathway or whether this type of pathway has never evolved in the worm. The existence of proteins related to Toll/IL-1R, Pelle/IRAK, Cactus/I-κB and TRAF argues not necessarily for the presence of a Rel/NF-κB-related pathway in *C. elegans*, as all these proteins have also been implicated in the JNK pathway, which is functional in *C. elegans*. However, I-κB proteins seem to act exclusively as regulators of the NF-κB pathway making the presence of a putative homologue of I-κB in *C. elegans* (Pujol et al. 2001), a very puzzling fact. Future work concerning the functional role of the putative *C. elegans* I-κB homologue will help us to resolve this issue.

4
Mammalian TRAF4 and DTRAF1 Are Highly Conserved and Involved in JNK Activation

Mammalian TRAF4 and its *Drosophila* homologue DTRAF1 are phylogenetically highly conserved, but also comparably distinct from other members of the TRAF family (Wajant et al. 2001). The sequence identity between both proteins is 45%, whereas the closest TRAF4-related human TRAF protein shares only 26% sequence identity (Wajant et al. 2001). Moreover, the expression pattern of TRAF4 and DTRAF1 are very similar: both molecules are detectable throughout embryogenesis and are predominantly found in undifferentiated cells, in particular neuronal precursors or epithelial progenitor cells (Krajewska et al. 1998; Masson et al. 1998; Preiss et al. 2001). Thus, the high degree of sequence similarity between DTRAF1 and TRAF4 is also reflected by a partially conserved expression pattern, suggesting that these two TRAF proteins fulfill an ancient conserved function in the differentiation of vertebrate and invertebrate cells. The recent analyses of TRAF4-deficient mice revealed a neural tube closure defect as well as malformation of rib, sternum, the spinal column and the upper respiratory tract. The latter is responsible for the increase in pulmonary inflammation found in these mice (Regnier et al. 2002; Shiels et al. 2000). Although DTRAF1 can interact with Pelle, an involvement in *Drosophila* Toll signaling was not confirmed by microinjection experiments in *Drosophila* eggs (Zapata et al. 2000). Similarily, there is no evidence for a major role of TRAF4 in mammalian NF-κB signaling. However, like other TRAF proteins, DTRAF1, as well as mammalian TRAF4, have been implicated in the activation of the JNK pathway. So, DTRAF1 has been identified as a binding partner of Misshapen, a MAP4K involved in the activation of the JNK pathway in flies (Liu et al. 1999). Moreover, TRAF4 interacts with p47phox, a NAD(P)H oxidase adapter subunit, and both synergistically induce oxidant production and JNK activation (Xu et al. 2002). Remarkably, impairment of the JNK pathway in *Drosophila* prevents dorsal closure (Sluss et al. 1996), a phenotype that is closely related with the neural tube closure defect described in TRAF4-deficent mice (Regnier et al. 2002). Despite the high degree of sequence similarity and domain architecture, there is one striking difference between TRAF4 and its *Drosophila* counterpart: in contrast to TRAF4, DTRAF1 has no amino-terminal RING domain (Liu et al. 1999; Preiss et al. 2001). Although the RING domain of TRAF6/DTRAF2 and TRAF2 have been implicated in the binding of effector proteins (Baud et al. 1999; Devin et al. 2001) and in the E3 ligase activity of these proteins (Shen et al. 2001; Brown et al. 2002), the relevance of this difference for TRAF4/DTRAF1-related functions has not been addressed yet.

Acknowledgements. This work was supported by Deutsche Forschungsgemeinschaft Grants Wa 1025/3-1 and Sonderforschungsbereich 495 project A5 and Deutsche Krebshilfe Grant 10–1751-Wa 3.

Appendix 1
Rel/NF-κB and I-κB Proteins

The nuclear factor kappa B (NF-κB)/Rel protein family comprises a group of phylogenetically conserved inducible transcription factors that act as homo- and/or heterodimers. Activation of Rel/NF-κB proteins can be induced by a huge variety of inducers, including cytokines, e.g. TNF and IL-1, bacterial products, UV radiation and reactive oxygen species. In non-stimulated cells, the Rel/NF-κB proteins are sequestered in the cytoplasm by interaction with proteins of the inhibitor of κB (I-κB) family. The NF-κB/Rel proteins are characterized by the conserved Rel homology domain (RHD) which mediates dimerization, DNA binding, nuclear localization and interaction with their inhibitory counterparts of the I-κB family. In mammals, five members of the NF-κB/Rel family have been described: RelA/p65, RelB, and c-Rel as well as NF-κB1/p50, which is released from its precursor p105 by a constitutively active proteolytic process, and NF-κB2/p52, which is released from its precursor p100 by induced proteolysis. In *Drosophila*, three Rel/NF-κB are known: Dorsal, Dif and Relish, the latter undergoes inducible proteolysis like NF-κB2. The I-κB proteins are defined by presence of six to seven copies of conserved helical protein domains, the so-called ankyrin repeats. In mammals, the I-κB family is represented by I-κBα, I-κBβ, I-κBγ, I-κBε, Bcl-3 and the aforementioned precursors of NF-κB1 and NF-κB2, p105 and p100. In *Drosophila*, there are two I-κB proteins, Cactus, and the precursor of Relish which carries six ankyrin repeats in its carboxy-terminal part (Dushay et al. 1996). The key event in activation of Rel/NF-κB transcription factors is the release of these proteins from their inhibitory I-κB counterparts. This release is based on the induced proteolytic degradation of I-κB proteins. In the case of the Rel/NF-κB precursors p100, p105 and Relish, proteolytic action is restricted to the carboxy-terminal ankyrin repeat-containing domain and therefore leads to the release of the active amino-terminal RHD. Proteolytic degradation of I-κB proteins solely consisting of ankyrin repeats as well as the proteolytic processing of RHD – ankyrin repeat chimeric precursors are triggered by phosphorylation of the I-κB proteins by a multi-protein complex designated as I-κB kinase (IKK) complex. The prototypic IKK complex in mammals consists of two related kinases, IKK1 and IKK2, phosphorylating I-κB proteins, a regulatory subunit called NEMO/IKKγ and Hsp90 and cdc37 (Chen et al. 2002). There is considerable evidence suggesting the existence of several distinct IKK complexes. This is for example obvious from the phenotype of IKK2-deficient mice, which are still able to form a cytokine-responsive IKK complex, but lack an IKK complex with the ability to promote phosphorylation-dependent processing of the NF-κB2 precursor p100. Moreover, a PMA-inducible I-κB kinase complex has been biochemically defined which lacks IKK1, IKK2 and NEMO, but contains a novel IKK1/2-related kinase, IKKε (Peters et al. 2000). As homologues of IKK2, NEMO/IKKγ and IKKε have also been described in *Drosophila*, distinct IKK complexes may also exist in this organism.

For more extended reviews concerning Rel/NF-κB and I-κB proteins, see Silverman and Maniatis (2001); Ghosh and Karin (2002); Karin and Lin (2002).

Appendix 2
MAP Kinase Cascades

Mitogen-activated protein kinase (MAPK) cascades are phylogenetically conserved signaling modules composed of three sequentially acting kinases: a MAPK, which is activated through threonine and tyrosine phosphorylation by a dual specificity MAPK kinase (MAPKK, MEK, MKK) which in turn becomes activated by a serine/threonine MAPKK kinase (MAPKKK, MEKK). MAPK cascades can be triggered by a huge variety of signals including growth factors, cytokines and physical cues like osmolarity. Consequently, they are critically involved in an extensive number of cellular processes. Three related groups of MAPKs are found from flies to mammals, defining functionally distinct MAPK modules: the extracellular signal-regulated kinases pathway, the c-Jun N-terminal kinases (JNK) pathway and the p38 kinases pathway. While the MAPKs are activated specifically by only one or two MAPKKs, these intermediate kinases are promiscuously activated by the various known MAPKKK. In the activation of the MAPKKK, a plethora of molecules can be involved including GTP-binding proteins, members of the germinal center kinase family of kinases and TRAF adaptor proteins.

For more extended reviews concerning MAP kinase cascades, see Davis (2000); Chang and Karin (2001); Hazzalin and Mahadevan (2002); Shaulian and Karin (2002).

Appendix 3
The Death Domain

The death domain (DD) has been originally identified by deletion mutagenesis of the apoptosis-inducing receptor Fas as that part of the molecule which is necessary to transduce the apoptotic signal generated on this receptor. Subsequently, this domain has also been found in other death-inducing receptors, in a variety of cytosolic adaptor proteins and also some kinases. It is currently clear that death domains act as protein–protein interaction modules by homophilic DD–DD association. X-ray structure analyses of several death domains revealed that this domain consists of six antiparallel amphiphatic α-helices arranged in a conserved overall fold. Structural and sequence analyses revealed that the death effector domain (DED) and the caspase-recruitment domain (CARD), two other conserved protein–protein interaction domains implicated in apoptotic pathways, also adopt a fold similar to that of death domains. Remarkably, DDs, DEDs and CARDs interact specifically with members of their own subfamily. Although all three domains are originally identified in homophilic protein–protein interactions involved in apoptosis signaling, they can also be part of non-apoptotic signaling pathways. Recently, a fourth domain, the Pyrin domain, has been identified by bioinformatics, which might also adopt a death domain-related fold.

For more extended reviews concerning the role and function of death domain-containing proteins, see Fesik (2000); Martinon et al. (2001).

References

Adachi O, Kawai T, Takeda K, Matsumoto M, Tsutsui H, Sakagami M, Nakanishi K, Akira S (1998) Targeted disruption of the *MyD88* gene results in loss of IL-1- and IL-18-mediated function. Immunity 9:143–150

Akira S, Takeda K, Kaisho T (2001) Toll-like receptors: critical proteins linking innate and acquired immunity. Nat Immunol 2:675–680

Bacher S, Grosshans J, Droge W, Schmitz ML (2001) The *Drosophila* proteins Pelle and Tube induce JNK/AP-1 activity in mammalian cells. FEBS Lett 497:153–158

Baud V, Liu ZG, Bennett B, Suzuki N, Xia Y, Karin M (1999) Signaling by proinflammatory cytokines: oligomerization of TRAF2 and TRAF6 is sufficient for JNK and IKK activation and target gene induction via an amino-terminal effector domain. Genes Dev 13:1297–1308

Belvin MP, Anderson KV (1996) A conserved signaling pathway: the *Drosophila* Toll-Dorsal pathway. Annu Rev Cell Dev Biol 12:393–416

Bodmer JL, Holler N, Reynard S, Vinciguerra P, Schneider P, Juo P, Blenis J, Tschopp J (2000) TRAIL receptor-2 signals apoptosis through FADD and caspase-8. Nat Cell Biol 2:241–243

Bonnard M, Mirtsos C, Suzuki S, Graham K, Huang J, Ng M, Itie A, Wakeham A, Shahinian A, Henzel WJ, Elia AJ, Shillinglaw W, Mak TW, Cao Z, Yeh WC (2000) Deficiency of T2 K leads to apoptotic liver degeneration and impaired NF-kappaB-dependent gene transcription. EMBO J 19:4976–4985

Brown KD, Hostager BS, Bishop GA (2002) Regulation of TRAF2 signaling by self-induced degradation. J Biol Chem 277:19433–19438

Burns K, Martinon F, Esslinger C, Pahl H, Schneider P, Bodmer JL, Di Marco F, French L, Tschopp J (1998) MyD88, an adapter protein involved in interleukin-1 signaling. J Biol Chem 273: 12203–12209

Cao Z, Henzel WJ, Gao X (1996) IRAK: a kinase associated with the interleukin-1 receptor. Science 271:1128–1131

Chang L, Karin M (2001) Mammalian MAP kinase signalling cascades. Nature 410:37–40

Chen G, Cao P, Goeddel DV (2002) TNF-induced recruitment and activation of the IKK complex require Cdc37 and Hsp90. Mol Cell 9:401–410

Chen P, Rodriguez A, Erskine R, Thach T, Abrams JM (1998) Dredd, a novel effector of the apoptosis activators reaper, grim, and hid in *Drosophila*. Dev Biol 201:202–216

Choe KM, Werner T, Stoven S, Hultmark D, Anderson KV (2002) Requirement for a peptidoglycan recognition protein (PGRP) in Relish activation and antibacterial immune responses in *Drosophila*. Science 296:359–362

Chung JY, Park YC, Ye H, Wu H (2002) All TRAFs are not created equal: common and distinct molecular mechanisms of TRAF-mediated signal transduction. J Cell Sci 115:679–688

Davis RJ (2000) Signal transduction by the JNK group of MAP kinases. Cell 103:239–252

DeLotto Y, DeLotto R (1998) Proteolytic processing of the *Drosophila* Spatzle protein by Easter generates a dimeric NGF-like molecule with ventralising activity. Mech Dev 72:141–148

Deng L, Wang C, Spencer E, Yang L, Braun A, You J, Slaughter C, Pickart C, Chen ZJ (2000) Activation of the IkappaB kinase complex by TRAF6 requires a dimeric ubiquitin-conjugating enzyme complex and a unique polyubiquitin chain. Cell 103:351–361

Devin A, Lin Y, Yamaoka S, Li Z, Karin M, Liu Z (2001) The alpha and beta subunits of IkappaB kinase (IKK) mediate TRAF2-dependent IKK recruitment to tumor necrosis factor (TNF) receptor 1 in response to TNF. Mol Cell Biol 21:3986–3994

Dushay MS, Asling B, Hultmark D (1996) Origins of immunity: Relish, a compound Rel-like gene in the antibacterial defense of *Drosophila*. Proc Natl Acad Sci USA 93:10343–10347

Edwards DN, Towb P, Wasserman SA (1997) An activity-dependent network of interactions links the Rel protein Dorsal with its cytoplasmic regulators. Development 124:3855–3864

Fesik SW (2000) Insights into programmed cell death through structural biology. Cell 103: 273–282

Fitzgerald KA, Palsson-McDermott EM, Bowie AG, Jefferies CA, Mansell AS, Brady G, Brint E, Dunne A, Gray P, Harte MT, McMurray D, Smith DE, Sims JE, Bird TA, O'Neill LA (2001) Mal (MyD88-adapter-like) is required for Toll-like receptor-4 signal transduction. Nature 413: 78–83

Galindo RL, Edwards DN, Gillespie SK, Wasserman SA (1995) Interaction of the pelle kinase with the membrane-associated protein tube is required for transduction of the dorsoventral signal in *Drosophila* embryos. Development 121:2209–2218

Georgel P, Naitza S, Kappler C, Ferrandon D, Zachary D, Swimmer C, Kopczynski C, Duyk G, Reichhart JM, Hoffmann JA (2001) *Drosophila* immune deficiency (IMD) is a death domain protein that activates antibacterial defense and can promote apoptosis. Dev Cell 1:503–514

Ghosh S, Karin M (2002) Missing pieces in the NF-kappaB puzzle. Cell 109 [Suppl]:S81–S96

Gottar M, Gobert V, Michel T, Belvin M, Duyk G, Hoffmann JA, Ferrandon D, Royet J (2002) The *Drosophila* immune response against Gram-negative bacteria is mediated by a peptidoglycan recognition protein. Nature 416:640–644

Grech A, Quinn R, Srinivasan D, Badoux X, Brink R (2000) Complete structural characterisation of the mammalian and *Drosophila* TRAF genes: implications for TRAF evolution and the role of RING finger splice variants. Mol Immunol 37:721–734

Grosshans J, Bergmann A, Haffter P, Nusslein-Volhard C (1994) Activation of the kinase Pelle by Tube in the dorsoventral signal transduction pathway of *Drosophila* embryo. Nature 372: 563–566

Hashimoto C, Hudson KL, Anderson KV (1988) The *Toll* gene of *Drosophila*, required for dorsal-ventral embryonic polarity, appears to encode a transmembrane protein. Cell 52:269–279

Hazzalin CA, Mahadevan LC (2002) MAPK-regulated transcription: a continuously variable gene switch? Nat Rev Mol Cell Biol 3:30–40

Hedengren M, Asling B, Dushay MS, Ando I, Ekengren S, Wihlborg M, Hultmark D (1999) Relish, a central factor in the control of humoral but not cellular immunity in *Drosophila*. Mol Cell 4:827–837

Heusch M, Lin L, Geleziunas R, Greene WC (1999) The generation of NF-kappaB2 p52: mechanism and efficiency. Oncogene 18:6201–6208

Hoffmann JA, Reichhart JM (2002) *Drosophila* innate immunity: an evolutionary perspective. Nat Immunol 3:121–126

Horng T, Medzhitov R (2001) *Drosophila* MyD88 is an adapter in the Toll signaling pathway. Proc Natl Acad Sci USA 98:12654–12658

Kanakaraj P, Schafer PH, Cavender DE, Wu Y, Ngo K, Grealish PF, Wadsworth SA, Peterson PA, Siekierka JJ, Harris CA, Fung-Leung WP (1998) Interleukin (IL)-1 receptor-associated kinase (IRAK) requirement for optimal induction of multiple IL-1 signaling pathways and IL-6 production. J Exp Med 187:2073–2079

Karin M, Lin A (2002) NF-kappaB at the crossroads of life and death. Nat Immunol 3:221–227

Kawai T, Adachi O, Ogawa T, Takeda K, Akira S (1999) Unresponsiveness of MyD88-deficient mice to endotoxin. Immunity 11:115–122

Kim YS, Han SJ, Ryu JH, Choi KH, Hong YS, Chung YH, Perrot S, Raibaud A, Brey PT, Lee WJ (2000) Lipopolysaccharide-activated kinase, an essential component for the induction of the antimicrobial peptide genes in *Drosophila melanogaster* cells. J Biol Chem 275:2071–2079

Kopp E, Medzhitov R, Carothers J, Xiao C, Douglas I, Janeway CA, Ghosh S (1999) ECSIT is an evolutionarily conserved intermediate in the Toll/IL-1 signal transduction pathway. Genes Dev 13:2059–2071

Krajewska M, Krajewski S, Zapata JM, Van Arsdale T, Gascoyne RD, Berern K, McFadden D, Shabaik A, Hugh J, Reynolds A, Clevenger CV, Reed JC (1998) TRAF-4 expression in epithelial progenitor cells. Analysis in normal adult, fetal, and tumor tissues. Am J Pathol 152: 1549–1561

Lagueux M, Perrodou E, Levashina EA, Capovilla M, Hoffmann JA (2000) Constitutive expression of a complement-like protein in Toll and JAK gain-of-function mutants of *Drosophila*. Proc Natl Acad Sci USA 97:11427–11432

Lehrer RI, Ganz T (2002) Defensins of vertebrate animals. Curr Opin Immunol 14:96–102

Leitges M, Sanz L, Martin P, Duran A, Braun U, Garcia JF, Camacho F, Diaz-Meco MT, Rennert PD, Moscat J (2001) Targeted disruption of the zetaPKC gene results in the impairment of the NF-kappaB pathway. Mol Cell 8:771–780

Lemaitre B, Kromer-Metzger E, Michaut L, Nicolas E, Meister M, Georgel P, Reichhart JM, Hoffmann JA (1995) A recessive mutation, immune deficiency (imd), defines two distinct control pathways in the *Drosophila* host defense. Proc Natl Acad Sci USA 92:9465–9469

Lemaitre B, Nicolas E, Michaut L, Reichhart JM, Hoffmann JA (1996) The dorsoventral regulatory gene cassette spatzle/Toll/cactus controls the potent antifungal response in *Drosophila* adults. Cell 86:973–983

LeMosy EK, Hong CC, Hashimoto C (1999) Signal transduction by a protease cascade. Trends Cell Biol 9:102–107

Leulier F, Rodriguez A, Khush RS, Abrams JM, Lemaitre B (2000) The *Drosophila* caspase Dredd is required to resist Gram-negative bacterial infection. EMBO Rep 1:353–358

Levashina EA, Ohresser S, Lemaitre B, Imler JL (1998) Two distinct pathways can control expression of the gene encoding the *Drosophila* antimicrobial peptide metchnikowin. J Mol Biol 278: 515–527

Levashina EA, Langley E, Green C, Gubb D, Ashburner M, Hoffmann JA, Reichhart JM (1999) Constitutive activation of *toll*-mediated antifungal defense in serpin-deficient *Drosophila*. Science 285:1917–1919

Levashina EA, Moita LF, Blandin S, Vriend G, Lagueux M, Kafatos FC (2001) Conserved role of a complement-like protein in phagocytosis revealed by dsRNA knockout in cultured cells of the mosquito, *Anopheles gambiae*. Cell 104:709–718

Li S, Strelow A, Fontana EJ, Wesche H (2002) IRAK-4: a novel member of the IRAK family with the properties of an IRAK-kinase. Proc Natl Acad Sci USA 99:5567–5572

Li X, Commane M, Burns C, Vithalani K, Cao Z, Stark GR (1999) Mutant cells that do not respond to interleukin-1 (IL-1) reveal a novel role for IL-1 receptor-associated kinase. Mol Cell Biol 19:4643–4652

Li X, Commane M, Jiang Z, Stark GR (2001) IL-1-induced NF-kappa B and c-Jun N-terminal kinase (JNK) activation diverge at IL-1 receptor-associated kinase (IRAK). Proc Natl Acad Sci USA 98:4461–4465

Liu C, Xu Z, Gupta D, Dziarski R (2001) Peptidoglycan recognition proteins: a novel family of four human innate immunity pattern recognition molecules. J Biol Chem 276:34686–34694

Liu H, Su YC, Becker E, Treisman J, Skolnik EY (1999) A *Drosophila* TNF-receptor-associated factor (TRAF) binds the ste20 kinase Misshapen and activates Jun kinase. Curr Biol 9:101–104

Lomaga MA, Yeh WC, Sarosi I, Duncan GS, Furlonger C, Ho A, Morony S, Capparelli C, Van G, Kaufman S, van der HA, Itie A, Wakeham A, Khoo W, Sasaki T, Cao Z, Penninger JM, Paige CJ, Lacey DL, Dunstan CR, Boyle WJ, Goeddel DV, Mak TW (1999) TRAF6 deficiency results in osteopetrosis and defective interleukin-1, CD40, and LPS signaling. Genes Dev 13:1015–1024

Lu Y, Wu LP, Anderson KV (2001) The antibacterial arm of the *Drosophila* innate immune response requires an IkappaB kinase. Genes Dev 15:104–110

Manfruelli P, Reichhart JM, Steward R, Hoffmann JA, Lemaitre B (1999) A mosaic analysis in *Drosophila* fat body cells of the control of antimicrobial peptide genes by the Rel proteins Dorsal and DIF. EMBO J 18:3380–3391

Martinon F, Hofmann K, Tschopp J (2001) The pyrin domain: a possible member of the death domain-fold family implicated in apoptosis and inflammation. Curr Biol 11:R118–R120

Masson R, Regnier CH, Chenard MP, Wendling C, Mattei MG, Tomasetto C, Rio MC (1998) Tumor necrosis factor receptor associated factor 4 (TRAF4) expression pattern during mouse development. Mech Dev 71:187–191

McWhirter SM, Pullen SS, Holton JM, Crute JJ, Kehry MR, Alber T (1999) Crystallographic analysis of CD40 recognition and signaling by human TRAF2. Proc Natl Acad Sci USA 96:8408–8413

Medzhitov R, Preston-Hurlburt P, Kopp E, Stadlen A, Chen C, Ghosh S, Janeway CA Jr (1998) MyD88 is an adaptor protein in the hToll/IL-1 receptor family signaling pathways. Mol Cell 2:253–258

Meister M, Hetru C, Hoffmann JA (2000) The antimicrobial host defense of *Drosophila*. Curr Top Microbiol Immunol 248:17–36

Meng X, Khanuja BS, Ip YT (1999) Toll receptor-mediated *Drosophila* immune response requires Dif, an NF-kappaB factor. Genes Dev 13:792–797

Michel T, Reichhart JM, Hoffmann JA, Royet J (2001) *Drosophila* Toll is activated by Gram-positive bacteria through a circulating peptidoglycan recognition protein. Nature 414:756–759

Morisato D, Anderson KV (1994) The *spatzle* gene encodes a component of the extracellular signaling pathway establishing the dorsal-ventral pattern of the *Drosophila* embryo. Cell 76: 677–688

Muzio M, Ni J, Feng P, Dixit VM (1997) IRAK (Pelle) family member IRAK-2 and MyD88 as proximal mediators of IL-1 signaling. Science 278:1612–1615

Muzio M, Natoli G, Saccani S, Levrero M, Mantovani A (1998) The human *toll* signaling pathway: divergence of nuclear factor kappaB and JNK/SAPK activation upstream of tumor necrosis factor receptor-associated factor 6 (TRAF6). J Exp Med 187:2097–2101

Ni CZ, Welsh K, Leo E, Chiou C, Wu H, Reed JC, Ely KR (2000) Molecular basis for CD40 signaling mediated by TRAF3. Proc Natl Acad Sci USA 97:10395–10399

Ninomiya-Tsuji J, Kishimoto K, Hiyama A, Inoue J, Cao Z, Matsumoto K (1999) The kinase TAK1 can activate the NIK-I kappaB as well as the MAP kinase cascade in the IL-1 signalling pathway. Nature 398:252–256

Park YC, Burkitt V, Villa AR, Tong L, Wu H (1999) Structural basis for self-association and receptor recognition of human TRAF2. Nature 398:533–538

Peters RT, Liao SM, Maniatis T (2000) IKKepsilon is part of a novel PMA-inducible IkappaB kinase complex. Mol Cell 5:513–522

Pomerantz JL, Baltimore D (1999) NF-kappaB activation by a signaling complex containing TRAF2, TANK and TBK1, a novel IKK-related kinase. EMBO J 18:6694–6704

Preiss A, Johannes B, Nagel AC, Maier D, Peters N, Wajant H (2001) Dynamic expression of *Drosophila* TRAF1 during embryogenesis and larval development. Mech Dev 100:109–113

Pujol N, Link EM, Liu LX, Kurz CL, Alloing G, Tan MW, Ray KP, Solari R, Johnson CD, Ewbank JJ (2001) A reverse genetic analysis of components of the Toll signaling pathway in *Caenorhabditis elegans*. Curr Biol 11:809–821

Qian Y, Commane M, Ninomiya-Tsuji J, Matsumoto K, Li X (2001) IRAK-mediated translocation of TRAF6 and TAB2 in the interleukin-1-induced activation of NF-kappaB. J Biol Chem 276: 41661–41667

Ramet M, Manfruelli P, Pearson A, Mathey-Prevot B, Ezekowitz RA (2002) Functional genomic analysis of phagocytosis and identification of a *Drosophila* receptor for *E. coli*. Nature 416: 644–648

Regnier CH, Masson R, Kedinger V, Textoris J, Stoll I, Chenard MP, Dierich A, Tomasetto C, Rio MC (2002) Impaired neural tube closure, axial skeleton malformations, and tracheal ring disruption in TRAF4-deficient mice. Proc Natl Acad Sci USA 99:5585–5590

Rich T, Allen R, Trowsdale J (2000) How low can Toll go? Trends Genet 16:292–294

Rutschmann S, Jung AC, Hetru C, Reichhart JM, Hoffmann JA, Ferrandon D (2000a) The Rel protein DIF mediates the antifungal but not the antibacterial host defense in *Drosophila*. Immunity 12:569–580

Rutschmann S, Jung AC, Zhou R, Silverman N, Hoffmann JA, Ferrandon D (2000b) Role of *Drosophila* IKK gamma in a *toll*-independent antibacterial immune response. Nat Immunol 1:342–347

Sanz L, Diaz-Meco MT, Nakano H, Moscat J (2000) The atypical PKC-interacting protein p62 channels NF-kappaB activation by the IL-1-TRAF6 pathway. EMBO J 19:1576–1586

Schwandner R, Yamaguchi K, Cao Z (2000) Requirement of tumor necrosis factor receptor-associated factor (TRAF) 6 in interleukin 17 signal transduction. J Exp Med 191:1233–1240

Senftleben U, Cao Y, Xiao G, Greten FR, Krahn G, Bonizzi G, Chen Y, Hu Y, Fong A, Sun SC, Karin M (2001) Activation by IKKalpha of a second, evolutionary conserved, NF-kappa B signaling pathway. Science 293:1495–1499

Shaulian E and Karin M (2002) AP-1 as a regulator of cell life and death. Nat Cell Biol 4:E131–E136

Shen B, Manley JL (1998) Phosphorylation modulates direct interactions between the Toll receptor, Pelle kinase and Tube. Development 125:4719–4728

Shen B, Liu H, Skolnik EY, Manley JL (2001) Physical and functional interactions between *Drosophila* TRAF2 and Pelle kinase contribute to Dorsal activation. Proc Natl Acad Sci USA 98:8596–8601

Shiels H, Li X, Schumacker PT, Maltepe E, Padrid PA, Sperling A, Thompson CB, Lindsten T (2000) TRAF4 deficiency leads to tracheal malformation with resulting alterations in air flow to the lungs. Am J Pathol 157:679–688

Shimada T, Kawai T, Takeda K, Matsumoto M, Inoue J, Tatsumi Y, Kanamaru A, Akira S (1999) IKK-i, a novel lipopolysaccharide-inducible kinase that is related to IkappaB kinases. Int Immunol 11:1357–1362

Silverman N, Maniatis T (2001) NF-kappaB signaling pathways in mammalian and insect innate immunity. Genes Dev 15:2321–2342

Silverman N, Zhou R, Stoven S, Pandey N, Hultmark D, Maniatis T (2000) A *Drosophila* IkappaB kinase complex required for Relish cleavage and antibacterial immunity. Genes Dev 14: 2461–2471

Sluss HK, Han Z, Barrett T, Davis RJ, Ip YT (1996) A JNK signal transduction pathway that mediates morphogenesis and an immune response in *Drosophila*. Genes Dev 10:2745–2758

Song HY, Regnier CH, Kirschning CJ, Goeddel DV, Rothe M (1997) Tumor necrosis factor (TNF)-mediated kinase cascades: bifurcation of nuclear factor-kappaB and c-jun N-terminal kinase (JNK/SAPK) pathways at TNF receptor-associated factor 2. Proc Natl Acad Sci USA 94: 9792–9796

Spencer E, Jiang J, Chen ZJ (1999) Signal-induced ubiquitination of IkappaBalpha by the F-box protein Slimb/beta-TrCP. Genes Dev 13:284–294

Stoven S, Ando I, Kadalayil L, Engstrom Y, Hultmark D (2000) Activation of the *Drosophila* NF-kappaB factor Relish by rapid endoproteolytic cleavage. EMBO Rep 1:347–352

Suzuki N, Suzuki S, Duncan GS, Millar DG, Wada T, Mirtsos C, Takada H, Wakeham A, Itie A, Li S, Penninger JM, Wesche H, Ohashi PS, Mak TW, Yeh WC (2002) Severe impairment of interleukin-1 and Toll-like receptor signalling in mice lacking IRAK-4. Nature 416:750–756

Takaesu G, Ninomiya-Tsuji J, Kishida S, Li X, Stark GR, Matsumoto K (2001) Interleukin-1 (IL-1) receptor-associated kinase leads to activation of TAK1 by inducing TAB2 translocation in the IL-1 signaling pathway. Mol Cell Biol 21:2475–2484

Takayanagi H, Ogasawara K, Hida S, Chiba T, Murata S, Sato K, Takaoka A, Yokochi T, Oda H, Tanaka K, Nakamura K, Taniguchi T (2000) T-cell-mediated regulation of osteoclastogenesis by signalling cross-talk between RANKL and IFN-gamma. Nature 408:600–605

Takeuchi O, Kaufmann A, Grote K, Kawai T, Hoshino K, Morr M, Muhlradt PF, Akira S (2000) Cutting edge: preferentially the R-stereoisomer of the mycoplasmal lipopeptide macrophage-activating lipopeptide-2 activates immune cells through a Toll-like receptor 2- and MyD88-dependent signaling pathway. J Immunol 164:554–557

Tauszig-Delamasure S, Bilak H, Capovilla M, Hoffmann JA, Imler JL (2002) *Drosophila* MyD88 is required for the response to fungal- and Gram-positive bacterial infections. Nat Immunol 3:91–97

Tauszig S, Jouanguy E, Hoffmann JA, Imler JL (2000) Toll-related receptors and the control of antimicrobial peptide expression in *Drosophila*. Proc Natl Acad Sci USA 97:10520–10525

Tojima Y, Fujimoto A, Delhase M, Chen Y, Hatakeyama S, Nakayama K, Kaneko Y, Nimura Y, Motoyama N, Ikeda K, Karin M, Nakanishi M (2000) NAK is an IkappaB kinase-activating kinase. Nature 404:778–782

Towb P, Bergmann A, Wasserman SA (2001) The protein kinase Pelle mediates feedback regulation in the *Drosophila* Toll signaling pathway. Development 128:4729–4736

Vidal S, Khush RS, Leulier F, Tzou P, Nakamura M, Lemaitre B (2001) Mutations in the *Drosophila dTAK1* gene reveal a conserved function for MAPKKKs in the control of rel/NF-kappaB-dependent innate immune responses. Genes Dev 15:1900–1912

Wajant H, Scheurich P (2001) Tumor necrosis factor receptor-associated factor (TRAF) 2 and its role in TNF signaling. Int J Biochem Cell Biol 33:19–32

Wajant H, Muhlenbeck F, Scheurich P (1998) Identification of a TRAF (TNF receptor-associated factor) gene in *Caenorhabditis elegans*. J Mol Evol 47:656–662

Wajant H, Henkler F, Scheurich P (2001) The TNF-receptor-associated factor family. Scaffold molecules for cytokine receptors, kinases and their regulators. Cell Signal 13:389–400

Wang C, Deng L, Hong M, Akkaraju GR, Inoue J, Chen ZJ (2001) TAK1 is a ubiquitin-dependent kinase of MKK and IKK. Nature 412:346–351

Werner T, Liu G, Kang D, Ekengren S, Steiner H, Hultmark D (2000) A family of peptidoglycan recognition proteins in the fruit fly *Drosophila melanogaster*. Proc Natl Acad Sci USA 97: 13772–13777

Wesche H, Henzel WJ, Shillinglaw W, Li S, Cao Z (1997) MyD88: an adapter that recruits IRAK to the IL-1 receptor complex. Immunity 7:837–847

Wesche H, Gao X, Li X, Kirschning CJ, Stark GR, Cao Z (1999) IRAK-M is a novel member of the Pelle/interleukin-1 receptor-associated kinase (IRAK) family. J Biol Chem 274:19403–19410

Xiao G, Cvijic ME, Fong A, Harhaj EW, Uhlik MT, Waterfield M, Sun SC (2001a) Retroviral oncoprotein Tax induces processing of NF-kappaB2/p100 in T cells: evidence for the involvement of IKKalpha. EMBO J 20:6805–6815

Xiao G, Harhaj EW, Sun SC (2001b) NF-kappaB-inducing kinase regulates the processing of NF-kappaB2 p100. Mol Cell 7:401–409

Xu YC, Wu RF, Gu Y, Yang YS, Yang MC, Nwariaku FE, Terada LS (2002) Involvement of TRAF4 in oxidative activation of c-Jun amino terminal kinase. J Biol Chem 277:28051–28057

Yamin TT, Miller DK (1997) The interleukin-1 receptor-associated kinase is degraded by proteasomes following its phosphorylation. J Biol Chem 272:21540–21547

Yang D, Chertov O, Oppenheim JJ (2001) The role of mammalian antimicrobial peptides and proteins in awakening of innate host defenses and adaptive immunity. Cell Mol Life Sci 58: 978–989

Ye H, Park YC, Kreishman M, Kieff E, Wu H (1999) The structural basis for the recognition of diverse receptor sequences by TRAF2. Mol Cell 4:321–330

Zapata JM, Matsuzawa S, Godzik A, Leo E, Wasserman SA, Reed JC (2000) The *Drosophila* tumor necrosis factor receptor-associated factor-1 (DTRAF1) interacts with Pelle and regulates NF-kappaB activity. J Biol Chem 275:12102–12107

Zapata JM, Pawlowski K, Haas E, Ware CF, Godzik A, Reed JC (2001) A diverse family of proteins containing tumor necrosis factor receptor-associated factor domains. J Biol Chem 276: 24242–24252

Regulation of BMP and Activin Signaling in *Drosophila*

L. Parker, D.G. Stathakis, and K. Arora

Abstract. Cytokines of the TGF-β superfamily act through an evolutionarily conserved signaling pathway to elicit a diverse range of biological responses in vertebrates as well as invertebrates. *Drosophila* has proved to be a powerful system to unravel the profound complexities underlying the regulation of this superficially simple signaling system for two reasons – the availability of sophisticated genetic tools and the restricted number of core signaling components compared to vertebrates. A BMP signaling pathway in *Drosophila* that regulates growth, differentiation and morphogenesis of the embryo and the larva has been extensively characterized. This work has provided major insights into how gradients of secreted proteins can be established and maintained in vivo, allowing a single ligand to induce multiple cell fates rather than function as an on-off switch. More recently, an activin signaling pathway has also been delineated that is required for growth and neuronal function during development. This review provides an overview of TGF-β signaling in *Drosophila* with emphasis on the extensive modulation of signaling activity both within and outside the cell, that enables ligands to trigger specific and context-dependent effects.

1 Introduction

Ligands belonging to the TGF-β superfamily of signaling proteins have been identified in animals across the phylogenetic spectrum where they play diverse roles. These include regulation of the cell cycle and the immune response, axial patterning, cell fate specification, apoptosis and morphogenesis during development. The mature ligands are secreted as disulfide-linked dimers that are derived from the carboxy-terminal region of precursor proteins after cleavage of the prodomain. More than 40 ligands have been identified from vertebrates and invertebrates that can be classified into three major subgroups – the transforming growth factor-βs (TGF-βs), bone morphogenetic proteins (BMPs) and activins – based on amino acid homology within the ligand domain. The basic mechanisms involved in TGF-β signal transduction show a high degree of evolutionary conservation and therefore studies in model systems have proven to be a powerful means of identifying signaling components and defining their functional roles in vivo.

The essential framework of the signaling pathway is deceptively simple. At the extracellular surface, TGF-β-related ligands bind to a heteromeric receptor

Department of Developmental and Cell Biology, University of California Irvine, 4215 McGaugh Hall, Irvine, California 92697–2300, USA

Progress in Molecular and Subcellular Biology
A. Beschin (Ed.) Invertebrate Cytokines

complex of type I and type II transmembrane serine-threonine kinases (Fig. 1). Ligand binding triggers association between the receptors, enabling the constitutively active type II kinase to phosphorylate the type I receptor within a specific domain termed the GS box. In the BMP pathway, the type I receptor is believed to be the primary determinant of ligand selectivity, recruiting the appropriate type II receptor to form a higher affinity complex. In contrast, in the TGF-β and Activin pathways, ligand specificity is directed primarily by the type II receptor. The activated type I receptor initiates pathway-specific signaling by phosphorylating a distinct member of the receptor-regulated Smad (R-Smad) family of cytosolic signal transduction proteins. Activated R-Smads associate with Smad4, a common Smad (Co-Smad) shared by all three signaling pathways. The heteromeric Smad/Co-Smad complex translocates to the nucleus and associates with DNA-binding cofactors to directly activate or repress target gene expression. Although Smad4 is obligately required for most cellular responses, in a few instances signaling has been reported to occur in the absence of the Co-Smad (Wisotzkey et al. 1998; Sirard et al. 2000). Both the R-Smads and Co-Smads contain highly conserved MH1 and MH2 domains separated by a linker region. The MH2 domain of R-Smads contains an additional SSXS motif that is the substrate for phosphorylation by the type I receptor kinase. A third set of Smads, the inhibitory Smads (I-Smads), lack the MH1 domain and act as antagonists to prevent R-Smad phosphorylation (reviewed in Derynck and Feng 1997; Massague 1998; Piek et al. 1999; Massague and Wotton 2000; Wrana and Attisano 2000; Miyazono et al. 2001; Moustakas et al. 2001).

The contrast between the wide spectrum of cellular and developmental responses attributed to TGF-β-related ligands and the fact that they signal through relatively few (and shared) core components, emphasizes the extensive regulation that is in place to ensure the correct biological response. The simplest level of control involves spatial and temporal regulation of the transcription of core pathway components. Additional levels of regulatory hierarchy modulate the signaling output at each step in the pathway – extracellular, cytoplasmic and nuclear. The end result is precise control over the distribution and effective range of the ligand as well as the amplitude and duration of the signal in target cells. This is particularly important in the context of development where BMP and activin ligands are known to trigger distinct effects in a concentration-dependent manner rather than functioning as on-off switches (reviewed in Neumann and Cohen 1997; Podos and Ferguson 1999; Freeman and Gurdon 2002). Finally, transcription of the modulators is frequently regulated by the signal thus integrating an important feedback loop into the pathway. The sophisticated genetics and ability to manipulate gene activity in *Drosophila* has contributed significantly to establishing the importance of several key regulators of TGF-β signaling in vivo, and in determining their mechanism of action. Here, we review the *Drosophila* BMP and activin signaling pathways, with emphasis on the regulatory hierarchies that have emerged in recent years.

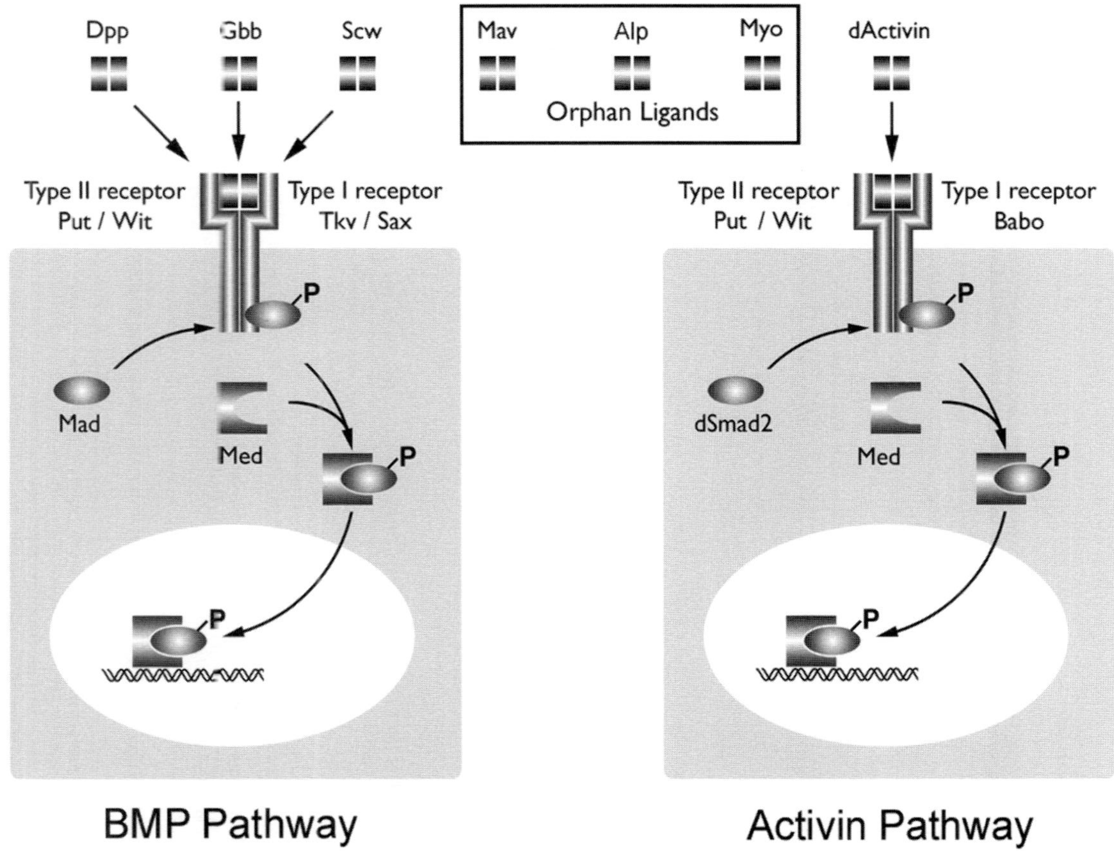

Fig. 1. The BMP and activin signaling pathways in *Drosophila*. The BMP ligands *Dpp*, *Gbb* and *Scw*, act through the type I receptors *Tkv* or *Sax*, resulting in phosphorylation of *Mad*, its association with the co-Smad *Med*, translocation of the complex into the nucleus, and regulation of target gene expression. The type II receptors *Put* and *Wit* display dual specificity and function in both the BMP and activin pathways. An activin pathway that involves the type I receptor *Babo* and *dSmad2* has been identified in *Drosophila*. Only one ligand, *dActivin*, has been shown to signal through *Babo* thus far. Phylogenetic analysis of *Mav*, *Alp*, and *Myo*, does not allow their assignment to a particular pathway

2 The BMP Ligands

A total of seven TGF-β-related ligands have been identified in *Drosophila* through a combination of molecular genetics and analysis of the complete genome sequence (Fig. 1). Three ligands clearly belong to the BMP subfamily, while the remaining four are more divergent (see Sect. 4).

Decapentaplegic (Dpp) represents the BMP2/4 ortholog in flies, while Screw (Scw) and Glass-bottom boat (Gbb/60A) are related to BMP 5/6/7/8 (Padgett et al. 1987; Wharton et al. 1991; Doctor et al. 1992; Arora et al. 1994).

2.1 *decapentaplegic*

The *dpp* gene is one of a small number of loci that are haplo-insufficient in *Drosophila*. Even a 50% reduction in levels of *dpp* activity leads to embryonic lethality, indicating the extreme sensitivity of the system to the ligand. As the name indicates (*decapentaplegic* or 15 defects), *dpp* mutants display a wide range of patterning abnormalities that are manifest in embryonic and adult tissues (Spencer et al. 1982). Genetic analysis has revealed a key role for Dpp as an instructive cue in coordinating patterning, growth and differentiation during development (Irish and Gelbart 1987). *dpp* mRNA is expressed in a dynamic spatial and temporal pattern reflective of the multiple requirements for the ligand. Three major *cis*-acting segments spanning ~55 kb of DNA have been delineated in the promoter – an upstream *dpp*-shv region required for expression in the embryonic gut and pupal wing, a central *dpp*-Hin region that directs embryonic expression, and a downstream *dpp*-disk region that regulates transcription in imaginal discs (St Johnston et al. 1990). These dispersed regulatory elements contribute to the genetic complexity of the locus and mediate *dpp* transcription in response to inputs from a variety of signaling pathways. Mutations in *dpp* that interfere with production or activity of the ligand affect all developmental events regulated by the gene, while mutations that disrupt specific enhancer elements affect *dpp* function in a tissue or stage specific manner. Among the processes that require *dpp* signaling are: oogenesis; specification of cell fates along the dorsal-ventral axis during embryogenesis; subdivision of the dorsal mesoderm; morphogenetic movements of dorsal closure; regional identity of the gut visceral mesoderm and endoderm; development of the heart, gastric cecae, salivary glands, and trachea; growth and patterning of imaginal discs.

One striking property of Dpp is its ability to function as a morphogen by inducing multiple cell fates in a concentration-dependent manner (Ferguson and Anderson 1992a, b; Wharton et al. 1993; Lecuit et al. 1996; Nellen et al. 1996). Understanding how stable gradients of signaling activity are established and interpreted is a fundamental issue in developmental biology. Two very different mechanisms have been invoked to explain how gradients of Dpp signaling can be generated. One is dependent on long range diffusion of the ligand from its site of synthesis, while the second involves diffusion of an extracellular inhibitor into the domain of ligand expression thus resulting in a gradient of signaling activity rather than ligand concentration. Dpp has been shown to act as a long range morphogen in the wing imaginal disc where it is expressed in a restricted manner but is required for growth and specification of cell fates in the entire wing primordium. The imaginal discs are small groups of epithelial cells set aside during embryogenesis, that grow and are patterned during larval and pupal stages and eventually differentiate to give rise to adult structures. *dpp* is transcribed in a narrow stripe of cells at the anterior-posterior compartment boundary of the wing disc, but can induce expression of target genes over a distance of 20 cell diameters. The Dpp-responsive genes *vestigial* (*vg*), *spalt* (*sal*) and *optomotor blind* (*omb*) are activated at different signaling thresholds and show nested domains of expression that further subdivide the wing disc into distinct regions (Lecuit et al. 1996; Nellen et al. 1996). An extracellular gradient of biologically active GFP-tagged Dpp has been visualized in the wing disc. Although Dpp-GFP can be seen at a distance of >30 cells in either direction from its site of synthesis, the gradient is unstable and spreads very rapidly (Entchev et al. 2000; Teleman and Cohen 2000). Furthermore, the secreted Dpp-GFP ligand is turned over rapidly and the shape of the extracellular Dpp gradient does not correlate well with intracellular activation of its signal transducer, the BMP-specific R-Smad Mothers against dpp (Mad). This has led to the idea that interaction of Dpp with its receptor Thickveins (Tkv) and internalization of the ligand/receptor complex may play a role in further stabilizing the gradient (Entchev et al. 2000). Although much of the Dpp-GFP is present at the cell surface, Entchev and coworkers have observed Dpp-GFP accumulation in endocytic vesicles and propose that a planar transcytosis mechanism may be involved in trafficking Dpp through cells and in controlling the shape of an intracellular gradient that defines cell fate.

A different mechanism for establishment of a morphogen gradient is encountered in the early embryo, where Dpp is required to induce distinct cell fates within the dorsal region. Peak levels of Dpp activity specify the dorsal most amnioserosa tissue, while lower levels are required for elaboration of the dorsal ectoderm. Reduction in Dpp signaling results in progressive loss of dorsal structures while increasing concentrations of Dpp can induce dorsal cell fates in a dosage-dependent manner, consistent with its role as a morphogen (Ferguson and Anderson 1992a; Wharton et al. 1993). Since *dpp* mRNA is expressed uniformly in the dorsal region, it is generally believed that a Dpp gradient is formed post-transcriptionally through modulation of ligand distribution or signaling activity. Extensive studies using biochemical and genetic assays have culminated in an elegant model in which the gradient of BMP signaling in dorsal embryonic cells is generated through the concerted action of three extracellular proteins – Short gastrulation (Sog), Twisted gastrulation

(Tsg) and Tolloid (Tld) – that modulate ligand activity in a graded manner (Shimell et al. 1991; Francois et al. 1994; Mason et al. 1994). The mechanism by which this occurs is described in Section 5.1.

2.2 *screw*

A second BMP ligand, Scw, also plays an essential role in early embryonic patterning and specification of dorsal cell fates (Arora et al. 1994). Mutations in *scw* result in partial ventralization of the embryo, compared to null *dpp* mutants that lack all dorsally derived structures and are completely ventralized. Functional analysis has established that the two ligands act in a qualitatively distinct manner. Excess Scw cannot rescue Dpp null embryos, thus Dpp is obligately required for dorsal patterning. In contrast, injection of *dpp* mRNA can restore dorsal cell fates in *scw* mutant embryos, indicating that Scw functions in a permissive manner and potentiates Dpp activity. These results indicate that both ligands act in a combinatorial manner to specify the complete range and extent of dorsal embryonic cell fates. The observation that *scw* is expressed exclusively in blastoderm stage embryos is intriguing, and points to a specific and temporally restricted requirement for Scw in promoting Dpp signaling (Arora et al. 1994).

2.3 *glass bottom boat*

A third BMP ligand, Gbb, was isolated in low stringency screens and subsequent analysis has established its role in specification of the embryonic gut and in growth and patterning of the imaginal discs (Chen et al. 1998; Khalsa et al. 1998; Wharton et al. 1999; Ray and Wharton 2001). *gbb* (like *scw*) acts cooperatively with *dpp* in some aspects of larval development, but can also function independently of Dpp in other contexts.

3 The BMP Signal Transduction Pathway

The limited number of ligands, receptors and Smads present in *Drosophila*, as well as the availability of versatile molecular and genetic tools, makes it a powerful system to examine specificity and cross-talk in the BMP and activin/TGF-β signaling pathways (Fig. 1). Only two type II receptors are found in *Drosophila*, the ActR-II homologue Punt (Put) and the BMPR-II homologue Wishful thinking (Wit) (Letsou et al. 1995; Ruberte et al. 1995; Aberle et al. 2002; Marques et al. 2002). Thus, all seven ligands must utilize either Put or Wit along with one of three type I receptors, Tkv, Saxophone (Sax) and Atr-1/Baboon (Babo), for signal transduction (Brummel et al. 1994; Nellen et al. 1994; Penton et al. 1994; Xie et al. 1994). Only two R-Smads, Mad and dSmad2, and a single Co-Smad, Medea (Med), are available to relay the signals downstream of these receptors. Mad and Med were identified in genetic screens for enhancers of a partial loss-of-function *dpp* mutant phenotype, and along with the *sma* genes in *Caenorhabditis elegans*, represent the founding members of the Smad family of signaling proteins (Raftery et al. 1995; Sekelsky et al. 1995; Newfeld et al. 1996; Savage et al. 1996; Chen et al. 1998; Das et al. 1998; Hudson et al. 1998; Wisotzkey et al. 1998). dSmad2 was identified by sequence homology and is most similar to vertebrate Smad2 that functions in activin signaling (Henderson and Andrew 1998).

Phenotypic analysis indicates that Put functions as the type II receptor for Dpp and Scw in early embryogenesis and in most (if not all) aspects of Dpp signaling at later stages of development (Fig. 1; Childs et al. 1993; Letsou et al. 1995; Ruberte et al. 1995). This is supported by biochemical studies in which Put was shown to bind Dpp. Interestingly, Put can also bind vertebrate activin in the presence of the appropriate type I receptor, suggesting that it is not a dedicated Dpp receptor. The type I receptor utilized by Dpp is Tkv (Brummel et al. 1994; Nellen et al. 1994; Penton et al. 1994), and the signal is transduced by Mad and Med (Raftery et al. 1995; Sekelsky et al. 1995; Newfeld et al. 1996; Das et al. 1998; Hudson et al. 1998; Wisotzkey et al. 1998). The binding specificity of Scw and Gbb has not yet been determined biochemically but their pathways have been delineated through functional assays. *scw* and *sax* have similar mutant phenotypes and Scw activity can be suppressed using a dominant-negative Sax receptor, suggesting that Sax functions in the Scw pathway (Neul and Ferguson 1998; Nguyen et al. 1998). Furthermore, mutations in *Mad* can ameliorate the phenotypes caused by overexpression of constitutively activated Sax (Das et al. 1998) and Sax specifically phosphorylates Mad in *Drosophila* S2 cell culture (D. Stathakis and K. Arora, unpubl.), indicating that Mad mediates Sax activity. Gbb was initially thought to signal through Sax in the wing imaginal disc, however subsequent studies have challenged that view and suggest that Gbb acts primarily through Tkv (Haerry et al. 1998; Ray and Wharton 2001). A comprehensive review of the core components of the BMP signaling pathway can be found in Raftery and Sutherland (1999).

The spatial and temporal pattern of the cellular response to BMP signaling in vivo can be visualized directly using antisera that specifically recognize the phosphorylated forms of Smad1/Mad. As expected from phenotypic studies, p-Mad accumulates in the dorsal region of the early embryo in a Dpp/Scw-dependent manner, as well as at other sites of Dpp signaling at later stages of embryonic and larval development (Fig. 2; Tanimoto et al. 2000; Dorfman and Shilo 2001). In addition, p-Mad accumulation is detected at novel sites where BMP signaling was not anticipated to play a role, such as in the nervous system (Fig. 2F). The relevance of these observations has been confirmed by recent studies that have uncovered a role for BMP signaling in controlling synaptic

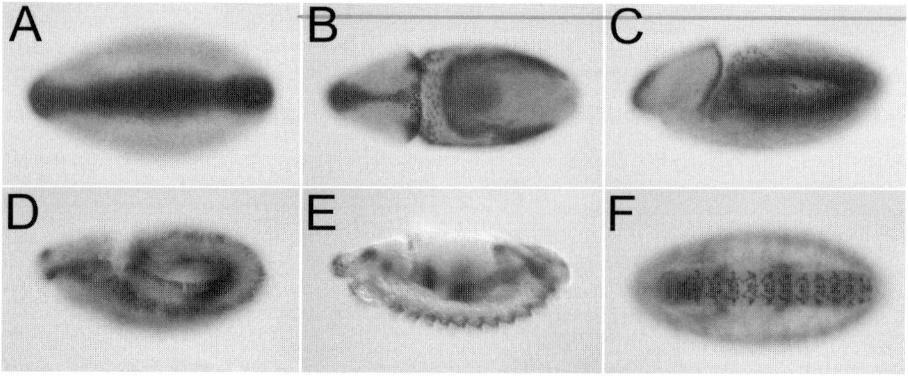

Fig. 2. Visualization of activated Mad during embryogenesis. Wild-type embryos stained with anti-PS1 antisera that recognize the phosphorylated form of Mad (p-Mad) reveal a dynamic spatial and temporal pattern of activation in response to endogenous BMP signaling. **A** At cellular blastoderm, a gradient of Dpp signaling results in peak levels of p-Mad in the mid-dorsal region with a sharp reduction in lateral cells. **B** In a stage 8 embryo, at germ-band extension, p-Mad, can be detected in dorsal cells in the anterior head region, the anterior transverse furrow, the amnioserosa and the dorsal epidermis (not seen in this view). **C** In stage 10 embryos, p-Mad is lost from the amnioserosa but remains enriched in the dorsal epidermis. **D** During germ-band retraction (stage 12), p-Mad expression refines and is found at the leading edge of the dorsal epidermis, as well as in a dorso-lateral stripe of epidermal cells. **E** At stage 13, Mad is activated in five discreet regions along the developing gut. **F** In a stage 15 embryo, a segmental-repeat pattern of p-Mad can be detected in a subset of cells in the ventral nerve cord. **A** and **B** are dorsal views, **C–E** are lateral views and **F** is a ventral view

growth and function at the neuromuscular junction. This aspect of BMP signaling is mediated by the type II receptor Wit and although the identity of the ligand has not been determined, circumstantial evidence points to the involvement of Gbb (Aberle et al. 2002; Marques et al. 2002).

Like other Smads, Mad and Med interact with a number of nuclear proteins that modulate the specificity of their transcriptional activity. These include the DNA-binding transcription factors Schnurri (Shn) as well as the coactivator CBP (Arora et al. 1995; Waltzer and Bienz 1999; Dai et al. 2000; Marty et al. 2000; Torres-Vazquez et al. 2000, 2001; Udagawa et al. 2000). The activity of these genes has only been investigated with respect to Dpp signaling and their potential involvement in other pathways remains to be determined.

4 The Activin Signaling Pathway and Non-Canonical Ligands

The activin/TGF-β signaling pathway in *Drosophila* is only just beginning to be characterized. The first evidence for such a pathway in flies came from the isolation of an activin type I receptor, Babo, that was shown to bind vertebrate activin in the presence of Put and phosphorylate dSmad2 in cell culture (Fig. 1; Brummel et al. 1999; Das et al. 1999). *babo* mutants die during larval stages and show an overall reduction in size, particularly noticeable in the brain. In addition, ectopic expression of constitutively active Babo in the wing disc results in increased cell numbers, suggesting that in *Drosophila*, activin signaling could be required for regulation of cell proliferation/growth. A fly activin homologue, dActivin, was identified based on sequence homology (Kutty et al. 1998). *dActivin* is expressed in the embryo and in the larva, primarily in the brain, ventral nerve cord (VNC) and the peripheral nervous system (Lo and Frasch 1999). Recently, activin signaling has been shown to be essential for neuronal remodeling during metamorphosis when the larval dendrites and axonal outgrowths are replaced by adult-specific processes (Zheng et al. 2003). These changes occur during the transition from larval to pupal stages and are coordinated by the steroid hormone ecdysone. It has been shown that loss of dActivin, Babo or dSmad2 blocks expression of a specific ecdysone receptor isoform in a subset of neurons in the mushroom body of the larval brain. As a result, these neurons are rendered insensitive to ecdysone and fail to remodel their synaptic arbors. Interestingly, the type II receptors Put and Wit act redundantly in this process, suggesting that both can transduce an activin signal (Fig. 1). This is supported by the observation that Wit can promote phosphorylation of dSmad2 in response to dActivin in cell culture (Zheng et al. 2003). These data emphasize that both Put and Wit can function as dual specificity type II receptors for BMP and activin ligands in vivo. However the two receptors cannot substitute for each other in all contexts, suggesting that some ligands may bind selectively to one receptor and not the other.

Three additional ligands have been identified with the completion of the *Drosophila* genome project. These are myoglianin (Myo), maverick (Mav) and activin-like-protein (Alp). All three ligands contain the nine cysteines characteristic of the TGF-β and activin subgroups. Myo is similar to mammalian GDF11 and myostatin/GDF8, that is highly expressed in embryonic and adult muscles and has been implicated in negative regulation of muscle mass (McPherron et al. 1997). In *Drosophila*, high levels of *myo* mRNA can be detected ubiquitously in the early embryo, while in late embryogenesis its expression is restricted to the muscle and glial precursor cells, as well as the larval brain (Lo and Frasch 1999). Although *myo* mutants have not been isolated and its receptor specificity is not known, the homology with vertebrate GDF8 and GDF11 suggests that *myo* may also utilize an activin/Smad2 pathway (Oh et al. 2002). Mav and Alp show low homology (<30%) to each other and all three subgroups of TGF-β superfamily, thus phylogenetic analysis and sequence comparison are not helpful in assigning these ligands to a particular pathway. *mav* is also expressed ubiquitously in the early embryo, and later in embryogenesis it can be detected in regions of the gut as well as in the visceral mesoderm that patterns the underlying gut endoderm (Nguyen et al. 2000). *alp* transcription is not initiated until late embryogenesis when the mRNA can be detected in a small number of cells in the VNC and the larval brain (Parker and Arora 2002). The absence of a prototypical TGF-β ligand in *Drosophila* suggests that BMPs and activins may represent the ancestral forms of this superfamily, while TGF-β ligands could be a more recent and perhaps chordate-specific evolutionary development. Consistent with this, the *C. elegans* genome also contains three BMP ligands but no TGF-β homologue (reviewed in Patterson and Padgett 2000), while BMP2/4 homologues have been isolated from the cnidarians, Hydra and Acropora (Hayward et al. 2002; B. Reinhardt and H. Bode, pers. comm.).

5 Extracellular Regulation of Ligand Activity

A large number of proteins have been identified in vertebrates and invertebrates that modulate signaling at the extracellular surface. These proteins are structurally diverse and utilize different mechanisms to regulate ligand activity suggesting that they may have evolved independently. Their activity results in alteration of ligand-receptor binding, ligand diffusion and/or the local ligand concentration. Extracellular modulators that have been identified and characterized in *Drosophila* are discussed below (Fig. 3).

5.1 Short Gastrulation, Twisted Gastrulation and Tolloid

The genes *sog*, *tsg* and *tld* were isolated in screens for mutations that disrupt embryonic dorso-ventral patterning. Molecular and functional characterization revealed that all three genes encode secreted proteins that act together to generate a gradient of Dpp/Scw signaling that is critical for specification of different dorsal cell fates in the embryo. Sog contains four cysteine-rich (CR) motifs and is homologous to vertebrate chordin that has been shown to antagonize BMP2/4 activity in *Xenopus* by forming a complex with the ligand, preventing it from interacting with the receptor (Francois et al. 1994; Sasai et al. 1995). In the embryo, *dpp* transcription is confined to dorsal cells while *scw* is transcribed ubiquitously. In contrast, *sog* is expressed in the ventro-lateral region of the embryo. Thus, Sog protein is expected to form a ventral to dorsal diffusion gradient that would result in an inverse gradient of BMP activity (Marques et al. 1997). Consistent with this, a Sog gradient that serves to limit Dpp signaling has been visualized in vivo (Srinivasan et al. 2002). Sog can discriminate between BMP ligands and has been shown to preferentially target Scw rather than Dpp activity, indicating that dorsal cell fates are specified by a gradient of Dpp and Scw signaling (Neul and Ferguson 1998; Nguyen et al. 1998). In *sog* mutant embryos, Dpp/Scw activity is not restricted and there

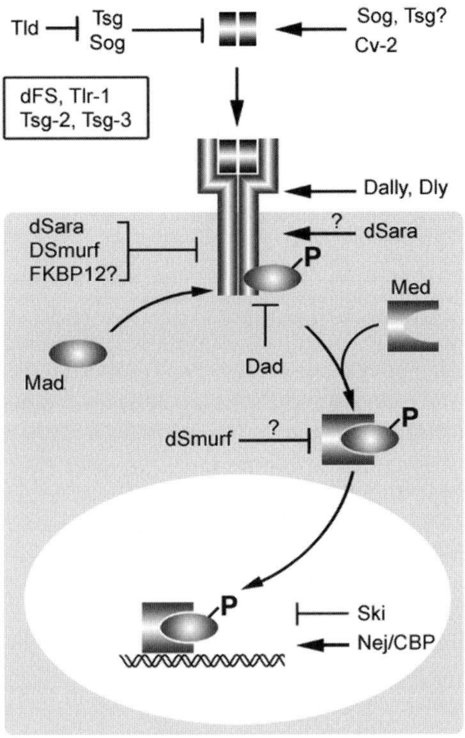

Fig. 3. Regulation of BMP signaling. The regulatory molecules that modulate BMP signaling and the levels at which they interact with the core signal-transduction pathway are shown. Positive inputs are depicted with *arrows* while inhibitory inputs are represented by a *bar*. Some factors (Sog and dSara) have both positive and negative effects on signaling depending on the context, and others (DSmurf) have been postulated to act at more than one level. The *boxed* proteins are secreted molecules that have been identified in *Drosophila* but whose specificity with respect to the BMP pathway remains to be determined. There is no functional data for the fly homologues of FKBP12 and Ski, and their putative roles are based on the activity ascribed to their vertebrate cognates. The role of most of these factors in activin signaling remains to be determined

is an expansion of dorso-lateral cell fates at the expense of ventral cell fates. Surprisingly, *sog* mutants also lack amnioserosa, the dorsal-most tissue that requires the highest levels of Dpp signaling. This unexpected finding suggested that Sog can also promote Dpp signaling in the extreme dorsal region of embryo. The dual affect of Sog on BMP signaling – short range inhibition and long range enhancement – has led to the idea that although Sog antagonizes ligand activity in ventral and lateral regions by forming a complex with the ligand, the complex itself could facilitate Dpp diffusion, perhaps by preventing ligand-receptor association (Marques et al. 1997; Ashe and Levine 1999). A metalloprotease Tld that is expressed only in dorsal cells can cleave ligand-associated Sog and release Dpp from an inactive complex. Thus, Tld prevents Sog from inhibiting BMP signaling in dorsal cells, and Sog in effect recaptures Dpp that has diffused ventrally and enables its release in dorsal regions. These studies highlight the fine interplay between the opposing activities of Sog and Tld that are crucial for achieving peak levels of Dpp/Scw signaling in the dorsal-most region of the embryo (Fig. 3). Although Tld is also secreted, its activity is restricted to the mid-dorsal region of the embryo through an as yet unidentified mechanism, ensuring that Dpp/Scw is preferentially liberated at this location. A number of studies have shown that although the dorsal-ventral axes in *Drosophila* and vertebrates are inverted relative to one another, they are specified through similar mechanisms. In *Xenopus*, BMP-4 promotes ventral development, while chordin promotes dorsal cell fates, and an amphibian homologue of Tld (Xolloid) has been shown to cleave chordin. Similar antagonistic interactions between homologous proteins are involved in patterning the dorsal-ventral axis in zebrafish and are likely to be important in other vertebrates as well (reviewed in Mullins 1998).

A third class of extracellular proteins that modulate BMP signaling is represented by Tsg, a protein that shows weak homology to human connective tissue growth factor (CTGF) (Mason et al. 1994). Tsg homologues have been found in *Xenopus*, zebrafish, mouse and humans but their exact role remains elusive. In *Drosophila*, *tsg* embryos show a phenotype very similar to that of *sog* mutants suggesting that Tsg may also have both BMP enhancing and inhibitory effects. In cell culture experiments, Tsg promotes the formation of a tripartite complex with Sog and Dpp that limits the effects of Dpp signaling (Ross et al. 2001). The zebrafish and *Xenopus* homologues act in a similar manner to block BMP activity (Chang et al. 2001; Scott et al. 2001). However, an independent study found that *Xenopus* Tsg can also enhance BMP signaling by promoting the dissociation of an inhibitory chordin fragment from the ligand (Oelgeschlager et al. 2000). Interestingly, these antagonists display differential specificity even with respect to closely related ligands and individual cysteine-rich (CR) domains retain the ability to bind and interfere with ligand function (reviewed in Garcia Abreu et al. 2002). In light of this, it is striking that the presence of Tsg alters the cleavage specificity of Tld for Sog to generate a novel fragment. Unlike full length Sog, a domain that mimics this novel fragment displays potent inhibitory activity with respect to Dpp (Yu et al. 2000). Thus, it is possible that different combinations of Sog- and Tsg-related proteins may have different effects on BMP signaling depending on the cellular context.

The mechanisms that modulate signaling in the embryo are likely to function in later development as well. A Tld-like protease, Tolkin/Tolloid-related-1 (Tlr1), mimics many aspects of *tld* expression in the embryo but is also transcribed in the nervous system and at additional sites in the larvae (Nguyen et al. 1994; Finelli et al. 1995). Mutations in *tlr1* are larval lethal and genetic analysis suggests that *tld* and *tlr1* do not have redundant functions. A potential role for *tlr1* in BMP signaling remains to be demonstrated. Two Tsg-related genes, Twisted gastrulation 2 (Tsg 2) and Twisted gastrulation 3 (Tsg 3) have been identified recently but are not yet functionally characterized (Vilmos et al. 2001). An additional Sog-related protein, Crossveinless-2 (Cv-2), is also present in *Drosophila*. Cv-2 contains five cysteine-rich (CR) domains and is required to potentiate Dpp or Gbb signaling in cells fated to give rise to the posterior cross-vein in the wing (Conley et al. 2000). The subtle phenotype of *cv-2* mutants illustrates how BMP signaling can be exquisitely regulated in vivo to precisely elaborate pattern in a small group of cells.

5.2
Dally and Dally-like

Another class of molecules that modulate BMP signaling at the extracellular level are the glypicans, that constitute a family of heparin sulphate proteoglycans tethered to the cell membrane by a glycosylphosphatidylinositol anchor (reviewed in Filmus and Selleck 2001). While several members of the glypican family exist in vertebrates, only two homologues, *division abnormally delayed* (*dally*) and *dally-like* (*dly*), are known in *Drosophila. dally* mutants were originally isolated on the basis of their effects on cell division in the developing *Drosophila* eye (Nakato et al. 1995). This phenotype, as well as defects in the antennae and genital discs, have been shown to result from a reduction in Dpp signaling (Jackson et al. 1997). However, *dally* and *dly* are not dedicated modifiers of BMP signaling and have also been implicated in the Wingless/Wnt signaling pathway (Lin and Perrimon 1999; Khare and Baumgartner 2000). The mechanism by which Dally regulates Dpp signaling is still unclear, but given its location at the cell surface it is postulated that it acts by altering the local distribution of ligand or its interaction with the receptor.

5.3
Follistatin and Follistatin-Related Protein (FSRP)

Like the BMPs, activin and TGF-β related ligands are also subject to regulation at the extracellular level. One well-known antagonist of activin signaling found in many vertebrate species is Follistatin, an extracellular protein that was isolated due its ability to inhibit secretion of follicle stimulating hormone. Follistatins typically contain a conserved amino-terminal domain followed by three closely related domains with a characteristic spacing of cysteine residues (Shimasaki et al. 1988). Two isoforms of Follistatin are known, both of which bind activin, although the shorter carboxy-truncated version is the more potent inhibitor (Inouye et al. 1991; Sugino et al. 1993). Binding of Follistatin to the ligand masks its ability to associate with the receptor, reminiscent of the mechanism by which Sog and Chordin function (de Winter et al. 1996; Phillips and de Kretser 1998). A role for Follistatin as an activin antagonist has been demonstrated in the vertebrate reproductive system and in *Xenopus* embryos (reviewed in Matzuk et al. 1996; Smith 1999). However, the specificity of Follistatin with respect to other ligands is still being elucidated. It is not clear that all of the defects encountered in Follistatin knock-out mice can be correlated with altered activin signaling (Matzuk et al. 1995), and there is some evidence that Follistatin can also bind BMPs and GDFs although with lower affinity than activin (Iemura et al. 1998; Gamer et al. 1999). A structural homologue, Follistatin related protein (FSRP), has been isolated that also binds activin and when overexpressed causes defects in the female reproductive system in mice (Schneyer et al. 2001). A *Drosophila* Follistatin homologue (dFS) was cloned recently (Haerry and O'Connor 2002) but its biological role has not yet been investigated. By analogy with the vertebrate protein, it might be expected to function as an inhibitor of dActivin and perhaps BMPs to a lesser extent. An FSRP homologue has not been identified in flies.

Although many of the extracellular regulators that modulate signaling by TGF-β-related ligands are common to vertebrates and *Drosophila*, there are a few for which no cognates exist. Notable amongst these are BMP antagonists of the Dan/Gremlin/Cerberus family of cysteine knot proteins initially identified in *Xenopus* (Hsu et al. 1998; Piccolo et al. 1999; Eimon and Harland 2001). Also missing are orthologs of the divergent ligands Lefty and Xenopus nodal related-3 that inhibit signaling by competing for the type II receptor. It is possible that these proteins are absent in *Drosophila* because they evolved after the bifurcation of the vertebrate and invertebrate lineages. Another potentially vertebrate-specific innovation is the regulation of TGF-β ligands through the formation of a latent complex, in which the prodomain remains associated with the secreted ligand in a non-covalent manner. Latent TGF-β binding protein (LTBP) is also implicated in this process and is involved in sequestering the ligand at the extracellular matrix (reviewed in Annes et al. 2003). *Drosophila* does not have a prototypical TGF-β ligand and so far there is no evidence for the existence of a LTBP.

6 Regulation of Receptor Signaling

Several molecules have been found to regulate the efficiency of receptor signaling. These include co-receptors that function at the level of the cell membrane and function by competing for/or promoting ligand binding. In addition at the cellular level, cytoplasmic proteins can promote receptor-Smad interaction, affect turnover of the receptor (or Smad), or compete for binding to the receptor active site (Fig. 3).

6.1 Synergistic Signaling by the Type I Receptors Tkv and Sax

As mentioned previously, the coordinate activity of two BMP ligands, Scw and Dpp, is required to specify the entire range of dorsal cell fates in early embryogenesis. This is an unusual example of co-dependence, where Dpp signaling is potentiated by Scw but the functional output of Scw signaling is entirely dependent on the Dpp pathway. Analysis of the non-reciprocal relationship between these two pathways has shown that modulation of Dpp signaling occurs through regulation of receptor activity (Neul and Ferguson 1998; Nguyen et al. 1998). Activation of the Scw type I receptor Sax has no phenotypic consequences in the absence of Dpp signaling but can elicit a strong synergistic response in the presence of a constitutively-active form of the Dpp receptor Tkv. While the exact mechanism for the synergy observed between Sax and Tkv is under investigation, current evidence indicates that the two pathways converge at the level of Mad phosphorylation and implicates the formation of heteromeric receptor complexes that signal more effectively (S. Park, D. Stathakis and K. Arora, unpubl.). Thus, synergy potentially functions to increase the efficiency of intracellular signaling or integrate signaling from different ligand-receptor combinations.

6.2 Smad Anchor for Receptor Activation

Efficient signaling by transmembrane receptors frequently requires pre-assembly of signal transduction components through the use of scaffolding proteins. The Smad anchor for receptor activation (SARA) is a FYVE domain protein that fulfills this role in TGF-β signaling (Tsukazaki et al. 1998). The FYVE motif is a lipid-binding domain that has been implicated in membrane association and could help localize SARA to the plasma membrane or endosomal vesicles. SARA was identified based on its ability to interact with Smad2/3 that mediate TGF-β and activin signaling, and subsequently shown to associate with the type I/type II receptor complex. Thus, it is believed to increase the effectiveness of R-Smad recruitment to the receptor. Importantly, phosphorylation of R-Smads allows their dissociation from SARA enabling efficient translocation of the activated Smad complex into the nucleus. The isolation of a *Drosophila* SARA homologue (dSara) in a yeast two-hybrid screen using the catalytic subunit of the type I serine-threonine phosphatase (PPIc), revealed an additional role for SARA-related proteins in BMP signaling (Bennett and Alphey 2002). Biochemical analysis showed that PPIc helps maintain low basal levels of receptor activation by dephosphorylating the Dpp receptor Tkv. Thus, dSara acts an adaptor between PPIc and Tkv. These results highlight the fact that SARA-related proteins could function as either positive or negative regulators of signaling depending on the proteins they recruit to the receptor.

6.3 Inhibitory Smads

Two mechanisms have been proposed to explain the inhibitory effect of the structurally distinct I-Smads that lack the amino terminal MH1 domain typical of R-Smads and Co-Smads. Several studies have shown that I-Smads can occupy the substrate-binding site of the type I receptor and block phosphorylation of the R-Smad (Hayashi et al. 1997; Imamura et al. 1997; Nakao et al. 1997). I-Smads have also been reported to compete with the Co-Smad for oligomerization with the R-Smad (Hata et al. 1998). I-Smad expression is induced in response to ligand activity, and thus they serve as a negative feedback loop to limit the duration of signaling. In vertebrates, the I-Smads are represented by Smad6 that preferentially inhibits BMP signaling, and Smad7 that is effective against TGF-β/activins as well as BMPs. A single I-Smad, Daughters against Dpp (Dad), exists in flies. The specificity of Dad has only been characterized with respect to Dpp signaling (Tsuneizumi et al. 1997). Clones of cells in the wing disc that lack Dad overexpress Dpp target genes, and ectopic Dad expression can suppress the effects of Dpp signaling. Dad has been shown to associate with Tkv but does not appear to bind Mad directly, thus arguing that it inhibits Mad indirectly by blocking its phosphorylation (Inoue et al. 1998).

6.4 Smad Ubiquitin Regulatory Factors

Two closely related Smad ubiquitin regulatory factors, Smurf1 and Smurf2, have been identified in vertebrates (Zhu et al. 1999; Lin et al. 2000; Zhang et al. 2001). These proteins contain HECT catalytic domains characteristic of E3 ubiquitin ligases that target proteins for degradation. Smurf1 has been shown to bind Smad1/5 implicating it in BMP signaling, while Smurf2 targets Smad1 and to some extent Smad2/3, potentially regulating both BMP and TGF-β/activin pathways. Extensive biochemical studies using cell culture and over-expression assays have outlined at least three different mechanisms through which Smurfs modulate TGF-β signaling. Smurfs can interact with R-Smads and regulate their turnover in a ligand-independent process, thus controlling their basal levels (Zhu et al. 1999; Lin et al. 2000; Zhang et al. 2001). A second mechanism involves association of Smurfs with the I-Smads (Smad 6/7) permitting the Smurfs to target the type I receptor for ubiquitination and subsequent degradation in the proteasome (Kavsak et al. 2000; Ebisawa et al. 2001). In yet a third mechanism, Smad2 has been shown to act as an adapter that enables Smurf2 to interact with and target the transcriptional corepressor SnoN (Bonni et al. 2001). The first two modes of Smurf action result in down regulation of signaling. In contrast, the third mechanism (degradation of SnoN) would be expected to enhance response to the ligand, since SnoN is a known repressor of TGF-β target gene expression. Given the opposing effects attributed to Smurf activity, it has been difficult to predict their in vivo function in vertebrates. The isolation of mutations in *Drosophila Smurf* (*DSmurf*) and analysis of the biological consequences of loss of Smurf activity has helped to fill this gap. Mutations in *DSmurf* were isolated based on their ability to suppress the embryonic patterning defects and lethality caused by reduction in Dpp activity, indicating that loss of *DSmurf* leads to an increase in Dpp signaling (Podos et al. 2001). Analysis of target gene expression in *DSmurf* mutants indicates that both the amplitude and the duration of the cellular response to Dpp is upregulated in these animals.

6.5 FKBP12

Mammalian FKBP12 was first characterized as a protein that binds FK506, an immunosuppressant drug that inhibits the calcium-activated phosphatase calcineurin. Subsequently, this ubiquitously expressed protein was shown to associate with the TGF-β type I receptor (TβRI) (Wang et al. 1994, 1996). TβRI mutants that cannot interact with FKBP12 show an increase in ligand-independent signaling (Chen et al. 1997). The functional relevance and in vivo requirement for FKBP12 in TGF-β signaling are still unresolved. One view is that FKBP12 binding to TβRI reduces basal levels of signaling in the cell that result from 'accidental' ligand-independent association between type I and type II receptors. An alternative hypothesis is that FKBP12 functions as a negative regulator of TβRI endocytosis (Yao et al. 2000). A *Drosophila* homologue of FKBP12 was isolated in a screen for differentially expressed genes in oogenesis and is transcribed in follicle cells overlying the dorsal anterior region of the oocyte (Bryant et al. 1999). Although no link has been established, it is worth noting that Dpp signaling is required for the development of dorsal anterior structures in the eggshell suggesting that dFKBP12 may be involved in modulating this pathway.

Not all regulatory strategies for controlling receptor activity encountered in vertebrates are represented in *Drosophila*. Striking amongst these is the absence of a pseudoreceptor resembling BAMBI. These are transmembrane proteins that are structurally similar to type I receptors in their extracellular domain but lack a carboxy-terminal kinase. This class of molecules have been shown to form stable complexes with both type I and type II receptors and down-regulate signaling by functioning as decoys (Onichtchouk et al. 1999; Tsang et al. 2000; Grotewold et al. 2001). It is likely that the greater number of ligands and the higher level of complexity in vertebrates provide strong evolutionary pressure for the development of additional regulatory strategies.

7 Nuclear Regulators of Signaling

A critical step in TGF-β signaling is the translocation of the R-Smad/Co-Smad complex to the nucleus where it directly regulates transcription of downstream target genes. Smads bind to DNA with weak affinity and recognize consensus sites that are of low complexity. It is now clear from a wide range of studies that association of Smads with other DNA-binding proteins is essential for pathway and promoter specific gene activation (reviewed in Massague and Wotton 2000). In addition to DNA binding, Smads also affect target gene expression by directly interacting with transcriptional coactivators and corepressors and recruiting them to the promoters. Furthermore, Smad activity itself is regulated through interaction with other factors that affect their turnover and nuclear accumulation. Smads have been shown to associate with a large number of proteins (>40) in a variety of organisms and cells, reflecting both the diversity of processes regulated by TGF-β-related ligands as well as the stringent control placed on Smad-dependent transcriptional activity (reviewed in Moustakas et al. 2001). It follows that some of these interactions are likely to be universal, while others may be organism or cell type specific (Fig. 3).

7.1 The Nuclear Repressor Brinker and the Smad-Interacting Factor Schnurri

The first direct evidence for the DNA-binding ability of Smads came from studies showing that *Drosophila* Mad recognizes distinct sequences in the disc-specific enhancer of the Dpp target gene *vg*. Mutations in these sites abolished Mad binding and Dpp responsiveness, indicating that Mad binding was critical for activation of *vg* expression (Kim et al. 1997). In light of this, an unexpected finding was that the transcriptional activation of a majority of *dpp* target genes involves a relief of repression mechanism that is dependent on interplay of Mad with two other DNA-binding proteins, Brinker (Brk) and Shn. Brk is a nuclear protein that functions as a sequence-specific transcriptional repressor of Dpp target genes. Mad negatively regulates *brk* transcription in response to Dpp signaling and thus *brk* is expressed throughout development in a dynamic pattern that is complementary to that of *dpp* (Campbell and Tomlinson 1999; Jazwinska et al. 1999a; 1999b; Minami et al. 1999). Overexpression of Dpp represses *brk* transcription, and loss of *brk* results in ectopic expression of *dpp* target genes like *sal* and *omb* (Fig. 4). Analysis of several Dpp responsive genes reveals that some (such as *sal*) require direct activation by Mad as well as relief of Brk repression for maximal levels of expression, while others (like *omb*) are activated entirely by downregulation of *brk*. Therefore, the opposing graded activities of Mad and Brk provide precise spatial control of gene expression in response to the Dpp morphogen. One mechanism by which Brk represses Dpp target genes is through direct competition with Mad for occupancy of enhancer elements since the two proteins have overlapping site specificities (Kirkpatrick et al. 2001; Rushlow et al. 2001). In other instances, Brk has been shown to interact with the corepressors CtBP and Groucho, suggesting that it can also repress transcription through independent binding sites (Ashe et al. 2000; Hasson et al. 2001).

Only one Smad-interacting factor, Shn, has been identified in *Drosophila* thus far. *shn* plays a role in Dpp signaling during late embryogenesis as well as in patterning and differentiation of adult structures (Arora et al. 1995; Grieder et al. 1995; Staehling-Hampton et al. 1995). It encodes a zinc finger DNA-binding protein that associates with Mad in a signaling dependent manner (Dai et al. 2000; Udagawa et al. 2000). In *shn* mutants, the expression of *dpp* target

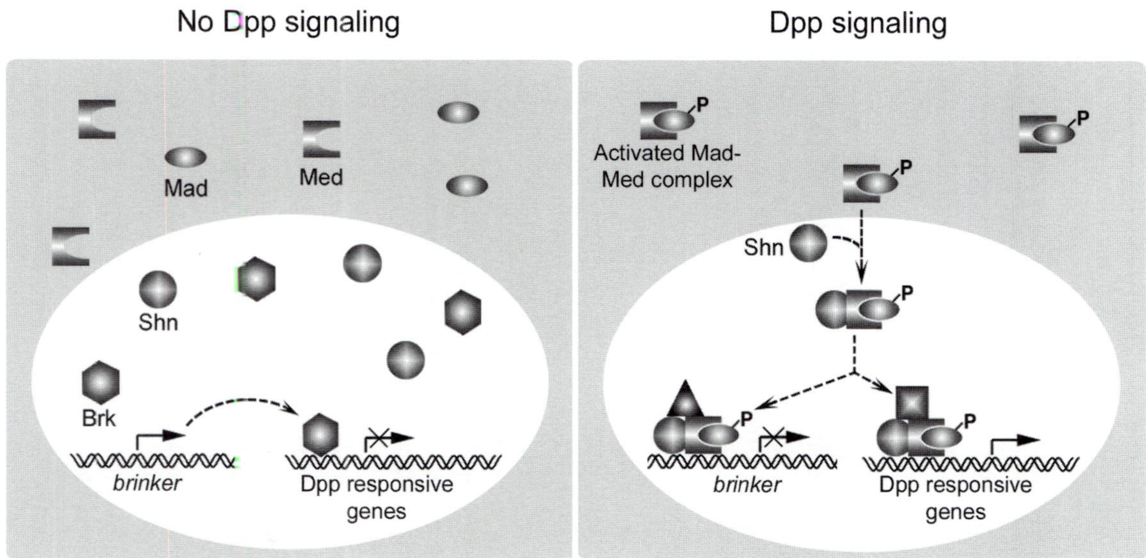

Fig. 4. Dual mechanism of Dpp target gene regulation. In the absence of Dpp signaling, the nuclear repressor Brk binds to sequences in the promoters of Dpp responsive genes preventing their expression. In response to Dpp signaling, an activated Mad/Med complex accumulates in the nucleus where it associates with Shn. This tripartite complex is involved in repression of *brk* transcription thus releasing Dpp responsive genes from inhibition. Some Dpp targets integrate both relief of repression by Brk as well as direct activation by Mad/Med for optimal levels of expression. The Shn/Mad/Med complex can participate in direct activation of Dpp target genes as well. It is likely that Shn facilitates Mad/Med activity by enhancing their interaction with DNA and/or components of the transcription machinery such as corepressors and coactivators in a promoter-specific manner

genes is lost and *brk* expression is upregulated. Two models have been proposed to explain how Shn functions (Fig. 4). One view is that Shn converts Mad into a repressor and is required exclusively to regulate *brk* expression (Marty et al. 2000). An alternative model argues that Shn is not a dedicated Mad corepressor and can participate in both aspects of Mad function. This view is supported by the fact that the Shn/Mad complex can directly activate transcription of the Dpp responsive gene *Ubx*, and that Shn is not redundant in the absence of Brk (Dai et al. 2000; Torres-Vazquez et al. 2001). Shn homologues have been identified in *C. elegans* as well as in vertebrates. While the worm homologue plays an essential role in the TGF-β/Sma pathway (R.W. Padgett, personal communication), the role of the three human Shn homologues in BMP signaling remains to be evaluated. Mouse knock-outs in one of them (MBP2/Shn2) blocks positive but not negative selection of thymocytes (Gascoigne 2001; Takagi et al. 2001). A second Shn homologue, kappa recognition component (KRC), is implicated in regulating gene activation in response to proinflammatory stimuli (Oukka et al. 2002). The involvement of Shn homologues in the immune system is interesting given that TGF-β signaling is known to modulate the inflammatory and infectious immune response in vertebrates (reviewed in Beschin et al. 2001). A parallel role for TGF-β-related ligands in the *Drosophila* immune response has not yet come to light.

Homologues of the vertebrate pathway-specific Smad-interacting factors, FAST-1 (forkhead activin signal transduer) and OAZ (olfactory factor O/E-1-associated zinc finger protein), have not been identified in *Drosophila* (Chen et al. 1996; Hata et al. 2000). However, the homeodomain protein Tinman (Tin) has been shown to bind its own promoter as well as the cardiac-specific enhancer of the *even-skipped* gene to mediate Dpp responsive expression in the mesoderm. In both these enhancers, the Tin sites are adjacent to Mad/Med sites and binding of all three proteins is required for gene expression (Xu et al. 1998; Halfon et al. 2000). Although direct interactions between Tin and Mad have not yet been demonstrated, it is possible that Tin acts as a tissue-specific Smad cofactor in mesodermal cells. A similar role can be postulated for Labial (*Lab*) and its cofactors Extradenticle and Homothorax, since binding of all three proteins is required for Dpp responsive expression of *lab* in the endoderm (Grieder et al. 1997; Ryoo et al. 1999).

7.2
Coactivators and Corepressors

Early functional assays established that the carboxy-terminal MH2 domain of Smad1 was sufficient for trans-activation when fused to a heterologous Gal4 DNA binding domain (Liu et al. 1996). Consistent with this idea, the MH2 domains of vertebrate Smads 1–4 have been shown to mediate interaction with coactivators of the p300/CBP and p300-CBP-associating factor (P/CAF) families, thus recruiting the basal transcriptional machinery to the promoter.

Similar interactions have been observed between Mad and the *Drosophila* CBP homologue Nejire (Nej). Furthermore, mutations in *nej* compromise the ability of Dpp to induce transcription of target genes in several tissues, indicating that this interaction is relevant in vivo (Waltzer and Bienz 1999).

Smads are also known to associate with several corepressors to mediate transcriptional repression. The best characterized amongst these are the oncoproteins of the Ski/SnoN family (reviewed in Liu et al. 2001). Ski/SnoN are integral components of the cellular transcriptional machinery that promote histone deacetylation and repress transcription. While Ski/SnoN do not bind DNA themselves, they are recruited to the TGF-β-responsive enhancers through interactions with Smads 2/3/4. Since transcription of SnoN is induced in response to TGF-βs, it also forms part of a negative feedback loop that serves to limit TGF-β signaling (Stroschein et al. 1999). A Ski/SnoN homologue is present in *Drosophila* but has not been characterized at either the molecular or functional level.

8 Concluding Remarks

In recent years, a great deal of effort has focused on identifying the core components of the BMP and activin pathways in *Drosophila* and understanding their mode of action. This work has elucidated the mechanisms by which morphogen gradients can be established in different developmental contexts, identified the founding members of the Smad family of signal transducers, delineated an alternate model for transcriptional regulation by Smads, and highlighted the importance of signaling modulators in vivo. These insights have in turn led to new questions and research directions. One important problem that remains is how gradients of signaling activity are stabilized. Although the general principles for formation of a Dpp gradient in the embryo and the wing have become evident, the exact mechanism by which ligands are propagated across a field of cells remains hotly debated. In the embryo, the precise biochemical sequence of events that allow the modulators Tsg/Sog to promote long-range diffusion/activity of the ligand are yet to be detailed. In the wing disc, a graded distribution of ligand has been visualized both at the cell surface as well as in vesicles inside the cells. While it is clear that receptor-mediated endocytosis plays a role in establishing a gradient of Dpp activity, the extent to which diffusion of Dpp in the extracellular space versus intracellular transcytosis contributes to this gradient is not known. In addition, it will be important to determine whether argosome-like vesicles and actin-containing cytoplasmic extensions called cytonemes, implicated in ligand transport, play a role in propagating BMP signals in imaginal discs (Ramirez-Weber and Kornberg 1999; Greco et al. 2001). Careful analysis using genetics and cell biological approaches will be required to resolve these issues.

It is likely that the core signaling pathways and biological roles of dActivin, Alp, Myo and Mav will be resolved in the near future. These studies will provide insights into how seven *Drosophila* ligands acting through only three type I receptors and two R-Smads can elicit distinct transcriptional responses in vivo. One possibility is that the synergy observed between the type I receptors Sax and Tkv represents a general mechanism for increasing the amplitude or the range of the intracellular responses to different ligands. Similar combinatorial codes of type I receptors may be utilized to specify distinct outputs. It is also clear that additional Smad cofactors remain to be identified that provide tissue and stage specificity to the response. An important question is whether the relief of repression mechanism (characterized in the context of Dpp signaling) is a widely used paradigm for regulation of target gene expression in these pathways. It will also be interesting to determine whether this model applies to vertebrates as well. Given the central role for Brk in Dpp signaling, it is intriguing that no structural homologues have been identified in worms or in vertebrates. It is possible that an analogous relief of repression mechanism is present in other organisms but is mediated by a structurally distinct molecule. In contrast, Shn homologues have been found in worms and humans, and it will be important to resolve whether they function as dedicated repressors of a Brk-like gene or if they also directly activate BMP/TGF-β target gene expression in other systems. Furthermore, a detailed dissection of BMP and activin responsive promoters will be necessary to reveal the interplay between the repressors and activators that are necessary for threshold activation of different target genes. Finally, it will be important to determine whether the modulators of signaling that have so far been characterized only in the context of the BMP pathway, also impact the activities of dActivin and other ligands.

Acknowledgements. We would like to thank Rahul Warrior for critical comments on the manuscript, Arthur Lander, Ken Cho and Ira Blitz for many fruitful discussions, and B. Reinhardt, H. Bode and R. Padgett for sharing unpublished data. This work was supported by grants from the NIH, NSF and March of Dimes Birth Defects Foundation to Kavita Arora.

References

Aberle H, Haghighi AP, Fetter RD, McCabe BD, Magalhaes TR, Goodman CS (2002) *wishful thinking* encodes a BMP type II receptor that regulates synaptic growth in *Drosophila*. Neuron 33:545–558

Annes JP, Munger JS, Rifkin DB (2003) Making sense of latent TGF-beta activation. J Cell Sci 116:217–224

Arora K, Dai H, Kazuko SG, Jamal J, O'Connor MB, Letsou A, Warrior R (1995) The *Drosophila schnurri* gene acts in the Dpp/ TGF-β signaling pathway and encodes a transcription factor homologous to the human MBP family. Cell 81:781–790

Arora K, Levine M, O'Connor M (1994) The *screw* gene encodes a ubiquitously expressed member of the TGF-β family required for specification of dorsal cell fates in the *Drosophila* embryo. Genes Dev 8:2588–2601

Ashe HL, Levine M (1999) Local inhibition and long-range enhancement of Dpp signal transduction by Sog. Nature 398:427–431

Ashe HL, Mannervik M, Levine M (2000) Dpp signaling thresholds in the dorsal ectoderm of the *Drosophila* embryo. Development 127:3305–3312

Bennett D, Alphey L (2002) PP1 binds Sara and negatively regulates Dpp signaling in *Drosophila* melanogaster. Nat Genet 31:419–423

Beschin A, Bilej M, Torreele E, De Baetselier P (2001) On the existence of cytokines in invertebrates. Cell Mol Life Sci 58:801–814

Bonni S, Wang HR, Causing CG, Kavsak P, Stroschein SL, Luo K, Wrana JL (2001) TGF-beta induces assembly of a Smad2-Smurf2 ubiquitin ligase complex that targets SnoN for degradation. Nat Cell Biol 3:587–595

Brummel TJ, Twombly V, Marques G, Wrana JL, Newfeld SJ, Attisano L, Massague J, O'Connor MB, Gelbart WM (1994) Characterization and relationship of Dpp receptors encoded by the *saxophone* and *thick veins* genes in *Drosophila*. Cell 78:252–261

Brummel TJ, Abdollah S, Haerry TE, Shimell MJ, Merriam J, Raftery L, Wrana JL, O'Connor MB (1999) The *Drosophila* activin receptor Baboon signals through dSmad2 and controls cell proliferation but not patterning during larval development. Genes Dev 13:98–111

Bryant Z, Subrahmanyan L, Tworoger M, LaTray L, Liu CR, Li MJ, van den Engh G, Ruohola-Baker H (1999) Characterization of differentially expressed genes in purified *Drosophila* follicle cells: toward a general strategy for cell type-specific developmental analysis. Proc Natl Acad Sci USA 96:5559–5564

Campbell G, Tomlinson A (1999) Transducing the Dpp morphogen gradient in the wing of *Drosophila*: regulation of Dpp targets by *brinker*. Cell 96:553–562

Chang C, Holtzman DA, Chau S, Chickering T, Woolf EA, Holmgren LM, Bodorova J, Gearing DP, Holmes WE, Brivanlou AH (2001) Twisted gastrulation can function as a BMP antagonist. Nature 410:483–487

Chen X, Rubock MJ, Whitman M (1996) A transcriptional partner for MAD proteins in TGF-beta signalling. Nature 383:691–696

Chen Y, Riese MJ, Killinger MA, Hoffmann FM (1998) A genetic screen for modifiers of *Drosophila decapentaplegic* signaling identifies mutations in *punt*, *Mothers against dpp* and the BMP-7 homologue, *60A*. Development 125:1759–1768

Chen YG, Liu F, Massague J (1997) Mechanism of TGF-beta receptor inhibition by FKBP12. EMBO J 16:3866–3876

Childs SR, Wrana JL, Arora K, Attisano L, O'Connor MB, Massague J (1993) Identification of a *Drosophila* activin receptor. Proc Natl Acad Sci USA 90:9475–9479

Conley CA, Silburn R, Singer MA, Ralston A, Rohwer-Nutter D, Olson DJ, Gelbart W, Blair SS (2000) Crossveinless 2 contains cysteine-rich domains and is required for high levels of BMP-like activity during the formation of the cross veins in *Drosophila*. Development 127:3947–3959

Dai H, Hogan C, Gopalakrishnan B, Torres-Vazquez J, Nguyen M, Park S, Raftery L A, Warrior R, Arora K (2000) The zinc finger protein Schnurri acts as a Smad partner in mediating the transcriptional response to Decapentaplegic. Dev Biol 227:373–387

Das P, Maduzia LL, Wang H, Finelli AL, Cho SH, Smith MM, Padgett RW (1998) The *Drosophila* gene *Medea* demonstrates the requirement for different classes of Smads in *dpp* signaling. Development 125:1519–1528

Das P, Inoue H, Baker JC, Beppu H, Kawabata M, Harland RM, Miyazono K, Padgett RW (1999) *Drosophila* dSmad2 and Atr-I transmit activin/TGF-beta signals. Genes Cells 4:123–134

de Winter JP, ten Dijke P, de Vries CJ, van Achterberg TA, Sugino H, de Waele P, Huylebroeck D, Verschueren K, van den Eijnden-van Raaij AJ (1996) Follistatins neutralize activin bioactivity by inhibition of activin binding to its type II receptors. Mol Cell Endocrinol 116:105–114

Derynck R, Feng XH (1997) TGF-beta receptor signaling. Biochim Biophys Acta 1333:F105–150

Doctor JS, Jackson PD, Raksha KE, Visalli M, Hoffmann FM (1992) Sequence, biochemical characterization, and developmental expression of a new member of the TGF-β superfamily in *Drosophila melanogaster*. Dev Biol 151:491–505

Dorfman R, Shilo BZ (2001) Biphasic activation of the BMP pathway patterns the *Drosophila* embryonic dorsal region. Development 128:965–972

Ebisawa T, Fukuchi M, Murakami G, Chiba T, Tanaka K, Imamura T, Miyazono K (2001) Smurf1 interacts with transforming growth factor-beta type I receptor through Smad7 and induces receptor degradation. J Biol Chem 276:12477–12480

Eimon PM, Harland RM (2001) *Xenopus* Dan, a member of the Dan gene family of BMP antagonists, is expressed in derivatives of the cranial and trunk neural crest. Mech Dev 107: 187–189

Entchev EV, Schwabedissen A, Gonzalez-Gaitan M (2000) Gradient formation of the TGF-beta homolog Dpp. Cell 103:981–991

Ferguson EL, Anderson KV (1992a) Decapentaplegic acts as a morphogen to organize dorsal-ventral pattern in the *Drosophila* embryo. Cell 71:451–461

Ferguson EL, Anderson KV (1992b) Localized enhancement and repression of the activity of the TGF-beta family member, *decapentaplegic*, is necessary for dorsal-ventral pattern formation in the *Drosophila* embryo. Development 114:583–597

Filmus J, Selleck SB (2001) Glypicans: proteoglycans with a surprise. J Clin Invest 108:497–501

Finelli AL, Xie T, Bossie CA, Blackman RK, Padgett RW (1995) The *tolkin* gene is a tolloid/BMP-1 homologue that is essential for *Drosophila* development. Genetics 141:271–281

Francois V, Solloway M, O'Neill JW, Emery J, Bier E (1994) Dorsal-ventral patterning of the *Drosophila* embryo depends on a putative negative growth factor encoded by the *short gastrulation* gene. Genes Dev 8:2602–2616

Freeman M, Gurdon JB (2002) Regulatory principles of developmental signaling. Annu Rev Cell Dev Biol 18:515–539

Gamer LW, Wolfman NM, Celeste AJ, Hattersley G, Hewick R, Rosen V (1999) A novel BMP expressed in developing mouse limb, spinal cord, and tail bud is a potent mesoderm inducer in *Xenopus* embryos. Dev Biol 208:222–232

Garcia Abreu J, Coffinier C, Larrain J, Oelgeschlager M, De Robertis EM (2002) Chordin-like CR domains and the regulation of evolutionarily conserved extracellular signaling systems. Gene 287:39–47

Gascoigne NR (2001) Positive selection in a Schnurri. Nat Immunol 2:989–991

Greco V, Hannus M, Eaton S (2001) Argosomes: a potential vehicle for the spread of morphogens through epithelia. Cell 106:633–645

Grieder NC, Nellen D, Burke R, Basler K, Affolter M (1995) *schnurri* is required for *Drosophila* Dpp signaling and encodes a zinc finger protein similar to the mammalian transcription factor PRDII-BF1. Cell 81:791–800

Grieder NC, Marty T, Ryoo HD, Mann RS, Affolter M (1997) Synergistic activation of a *Drosophila* enhancer by HOM/EXD and DPP signaling. EMBO J 16:7402–7410

Grotewold L, Plum M, Dildrop R, Peters T, Ruther U (2001) Bambi is coexpressed with Bmp-4 during mouse embryogenesis. Mech Dev 100:327–330

Haerry TE, Khalsa O, O'Connor MB, Wharton KA (1998) Synergistic signaling by two BMP ligands through the Sax and Tkv receptors controls wing growth and patterning in *Drosophila*. Development 125:3977–3987

Haerry TE, O'Connor MB (2002) Isolation of *Drosophila* activin and follistatin cDNAs using novel MACH amplification protocols. Gene 291:85–93

Halfon MS, Carmena A, Gisselbrecht S, Sackerson CM, Jimenez F, Baylies MK, Michelson AM (2000) Ras pathway specificity is determined by the integration of multiple signal-activated and tissue-restricted transcription factors. Cell 103:63–74

Hasson P, Muller B, Basler K, Paroush Z (2001) Brinker requires two corepressors for maximal and versatile repression in Dpp signaling. EMBO J 20:5725–5736

Hata A, Lagna G, Massague J, Hemmati-Brivanlou A (1998) Smad6 inhibits BMP/Smad1 signaling by specifically competing with the Smad4 tumor suppressor. Genes Dev 12:186–197

Hata A, Seoane J, Lagna G, Montalvo E, Hemmati-Brivanlou A, Massague J (2000) OAZ uses distinct DNA- and protein-binding zinc fingers in separate BMP- Smad and Olf signaling pathways. Cell 100:229–240

Hayashi H, Abdollah S, Qiu Y, Cai J, Xu Y, Grinnell B, Richardson M, Topper J, Gimbrone MJ, Wrana J, Falb D (1997) The MAD-related protein Smad7 associates with the TGF-beta receptor and functions as an antagonist of TGF-beta signaling. Cell 89:1165–1173

Hayward DC, Samuel G, Pontynen PC, Catmull J, Saint R, Miller DJ, Ball EE (2002) Localized expression of a dpp/BMP2/4 ortholog in a coral embryo. Proc Natl Acad Sci USA 99:8106–8111

Henderson KD, Andrew DJ (1998) Identification of a novel *Drosophila* SMAD on the X chromosome. Biochem Biophys Res Commun 252:195–201

Hsu DR, Economides AN, Wang X, Eimon PM, Harland RM (1998) The *Xenopus* dorsalizing factor Gremlin identifies a novel family of secreted proteins that antagonize BMP activities. Mol Cell 1:673–683

Hudson JB, Podos SD, Keith K, Simpson SL, Ferguson EL (1998) The *Drosophila Medea* gene is required downstream of *dpp* and encodes a functional homolog of human Smad4. Development 125:1407–1420

Iemura S, Yamamoto TS, Takagi C, Uchiyama H, Natsume T, Shimasaki S, Sugino H, Ueno N (1998) Direct binding of follistatin to a complex of bone-morphogenetic protein and its receptor inhibits ventral and epidermal cell fates in early Xenopus embryo. Proc Natl Acad Sci USA 95:9337–9342

Imamura T, Takase M, Nishihara A, Oeda E, Hanai J, Kawabata M, Miyazono K (1997) Smad6 inhibits signalling by the TGF-beta superfamily. Nature 389:622–626

Inoue H, Imamura T, Ishidou Y, Takase M, Udagawa Y, Oka Y, Tsuneizumi K, Tabata T, Miyazono K, Kawabata M (1998) Interplay of signal mediators of decapentaplegic (Dpp): molecular characterization of Mothers against dpp, Medea, and Daughters against dpp. Mol Biol Cell 9:2145–2156

Inouye S, Guo Y, DePaolo L, Shimonaka M, Ling N, Shimasaki S (1991) Recombinant expression of human follistatin with 315 and 288 amino acids: chemical and biological comparison with native porcine follistatin. Endocrinology 129:815–822

Irish VF, Gelbart WM (1987) The *decapentaplegic* gene is required for dorsal-ventral patterning of the *Drosophila* embryo. Genes Dev 1:868–879

Jackson SM, Nakato H, Sugiura M, Jannuzi A, Oakes R, Kaluza V, Golden C, Selleck SB (1997) Dally, a *Drosophila* glypican, controls cellular responses to the TGF-beta-related morphogen, Dpp. Development 124:4113–4120

Jazwinska A, Kirov N, Wieschaus E, Roth S, Rushlow C (1999a) The *Drosophila* gene *brinker* reveals a novel mechanism of Dpp target gene regulation. Cell 96:563–573

Jazwinska A, Rushlow C, Roth S (1999b) The role of *brinker* in mediating the graded response to Dpp in early *Drosophila* embryos. Development 126:3323–3334

Kavsak P, Rasmussen R K, Causing CG, Bonni S, Zhu H, Thomsen GH, Wrana JL (2000) Smad7 binds to Smurf2 to form an E3 ubiquitin ligase that targets the TGF beta receptor for degradation. Mol Cell 6:1365–1375

Khalsa O, Yoon JW, Torres-Schumann S, Wharton KA (1998) TGF-beta/BMP superfamily members, Gbb-60A and Dpp, cooperate to provide pattern information and establish cell identity in the *Drosophila* wing. Development 125:2723–2734

Khare N, Baumgartner S (2000) Dally-like protein, a new *Drosophila* glypican with expression overlapping with wingless. Mech Dev 99:199–202

Kim J, Johnson K, Chen HJ, Carroll S, Laughon A (1997) *Drosophila* Mad binds to DNA and directly mediates activation of *vestigial* by Decapentaplegic. Nature 388:304–308

Kirkpatrick H, Johnson K, Laughon A (2001) Repression of Dpp targets by binding of Brinker to Mad sites. J Biol Chem 276:18216–18222

Kutty G, Kutty RK, Samuel W, Duncan T, Jaworski C, Wiggert B (1998) Identification of a new member of transforming growth factor-beta superfamily in *Drosophila*: the first invertebrate activin gene. Biochem Biophys Res Commun 246:644–649

Lecuit T, Brook W J, Ng M, Calleja M, Sun H, Cohen SM (1996) Two distinct mechanisms for long-range patterning by Decapentaplegic in the *Drosophila* wing. Nature 381:387–393

Letsou A, Arora K, Wrana JL, Simin K, Twombly V, Jamal J, Staehling-Hampton K, Hoffmann FM, Gelbart WM, Massague J, et al. (1995) *Drosophila* Dpp signaling is mediated by the *punt* gene product: a dual ligand-binding type II receptor of the TGF beta receptor family. Cell 80: 899–908

Lin X, Liang M, Feng XH (2000) Smurf2 is a ubiquitin E3 ligase mediating proteasome-dependent degradation of Smad2 in transforming growth factor-beta signaling. J Biol Chem 275:36818–36822

Lin X, Perrimon N (1999) Dally cooperates with *Drosophila* Frizzled 2 to transduce Wingless signalling. Nature 400:281–284

Liu F, Hata A, Baker JC, Doody J, Carcamo J, Harland RM, Massague J (1996) A human Mad protein acting as a BMP-regulated transcriptional activator. Nature 381:620–623

Liu X, Sun Y, Weinberg RA, Lodish HF (2001) Ski/Sno and TGF-beta signaling. Cytokine Growth Factor Rev 12:1–8

Lo PC, Frasch M (1999) Sequence and expression of myoglianin, a novel *Drosophila* gene of the TGF-beta superfamily. Mech Dev 86:171–175

Marques G, Musacchio M, Shimell MJ, Wunnenberg-Stapleton K, Cho KW, O'Connor MB (1997) Production of a DPP activity gradient in the early *Drosophila* embryo through the opposing actions of the SOG and TLD proteins. Cell 91:417–426

Marques G, Bao H, Haerry TE, Shimell MJ, Duchek P, Zhang B, O'Connor MB (2002) The *Drosophila* BMP type II receptor Wishful Thinking regulates neuromuscular synapse morphology and function. Neuron 33:529–543

Marty T, Muller B, Basler K, Affolter M (2000) Schnurri mediates Dpp-dependent repression of *brinker* transcription. Nat Cell Biol 2:745–749

Mason ED, Konrad KD, Webb CD, Marsh JL (1994) Dorsal midline fate in *Drosophila* embryos requires *twisted gastrulation*, a gene encoding a secreted protein related to human connective tissue growth factor. Genes Dev 8:1489–1501

Massague J (1998) TGF-beta signal transduction. Annu Rev Biochem 67:753–791

Massague J, Wotton D (2000) Transcriptional control by the TGF-beta/Smad signaling system. EMBO J 19:1745–1754

Matzuk MM, Lu N, Vogel H, Sellheyer K, Roop DR, Bradley A (1995) Multiple defects and perinatal death in mice deficient in follistatin. Nature 374:360–363

Matzuk MM, Kumar TR, Shou W, Coerver KA, Lau AL, Behringer RR, Finegold MJ (1996) Transgenic models to study the roles of inhibins and activins in reproduction, oncogenesis, and development. Recent Prog Horm Res 51:123–154; discussion 155–157

McPherron AC, Lawler AM, Lee SJ (1997) Regulation of skeletal muscle mass in mice by a new TGF-beta superfamily member. Nature 387:83–90

Minami M, Kinoshita N, Kamoshida Y, Tanimoto H, Tabata T (1999) *brinker* is a target of Dpp in *Drosophila* that negatively regulates Dpp-dependent genes. Nature 398:242–246

Miyazono K, Kusanagi K, Inoue H (2001) Divergence and convergence of TGF-beta/BMP signaling. J Cell Physiol 187:265–276

Moustakas A, Souchelnytskyi S, Heldin CH (2001) Smad regulation in TGF-beta signal transduction. J Cell Sci 114:4359–4369

Mullins MC (1998) Holy Tolloido: Tolloid cleaves SOG/Chordin to free DPP/BMPs. Trends Genet 14:127–129

Nakato H, Futch TA, Selleck SB (1995) The *division abnormally delayed* (*dally*) gene: a putative integral membrane proteoglycan required for cell division patterning during postembryonic development of the nervous system in *Drosophila*. Development 121:3687–3702

Nakao A, Roijer E, Imamura T, Souchelnytskyi S, Stenman G, Heldin,C-H, ten Dijke P (1997) Identification of Smad 2, a Human Mad-related Protein in the Transforming Growth Factor Beta Signaling Pathway. J Biol Chem 272:2896–2900

Nellen D, Affolter M, Basler K (1994) Receptor Serine/Threonine kinases implicated in the control of *Drosophila* body pattern by *decapentaplegic*. Cell 78:225–237

Nellen D, Burke R, Struhl G, Basler K (1996) Direct and Long-Range Action of a DPP Morphogen Gradient. Cell 85:357–368

Neul JL, Ferguson EL (1998) Spatially restricted activation of the Sax receptor by Scw modulates Dpp/Tkv signaling in *Drosophila* dorsal-ventral patterning. Cell 95:483–494

Neumann C, Cohen S (1997) Morphogens and pattern formation. BioEssays 19:721–729

Newfeld SJ, Chartoff EH, Graff JM, Melton DA, Gelbart WM (1996) *Mothers against dpp* encodes a conserved cytoplasmic protein required in DPP/TGF-β responsive cells. Development 122: 1–10

Nguyen M, Park S, Marques G, Arora K (1998) Interpretation of a BMP activity gradient in *Drosophila* embryos depends on synergistic signaling by two type I receptors, SAX and TKV. Cell 95:495–506

Nguyen M, Parker L, Arora K (2000) Identification of *maverick*, a novel member of the TGF-beta superfamily in *Drosophila*. Mech Dev 95:201–206

Nguyen T, Jamal J, Shimell MJ, Arora K, O'Connor MB (1994) Characterization of *tolloid-related-1*: a BMP-1-like product that is required during larval and pupal stages of *Drosophila* development. Dev Biol 166:569–586

Oelgeschlager M, Larrain J, Geissert D, De Robertis EM (2000) The evolutionarily conserved BMP-binding protein Twisted gastrulation promotes BMP signalling. Nature 405:757–763

Oh SP, Yeo CY, Lee Y, Schrewe H, Whitman M, Li E (2002) Activin type IIA and IIB receptors mediate Gdf11 signaling in axial vertebral patterning. Genes Dev 16:2749–2754

Onichtchouk D, Chen YG, Dosch R, Gawantka V, Delius H, Massague J, Niehrs C (1999) Silencing of TGF-beta signalling by the pseudoreceptor BAMBI. Nature 401:480–485

Oukka M, Kim ST, Lugo G, Sun J, Wu LC, Glimcher LH (2002) A mammalian homolog of *Drosophila* schnurri, KRC, regulates TNF receptor-driven responses and interacts with TRAF2. Mol Cell 9:121–131

Padgett RW, Johnston RDS, Gelbart WM (1987) A transcript from a *Drosophila* pattern gene predicts a protein homologous to the transforming growth factor-(beta) family. Nature 325:81–84

Parker L, Arora K (2002) Activin signaling in *Drosophila*. Paper presented at: 43rd Annual *Drosophila* Resesearch Conference (San Diego).

Patterson GI, Padgett RW (2000) TGF beta-related pathways. Roles in *Caenorhabditis elegans* development. Trends Genet 16:27–33

Penton A, Chen Y, Staehling-Hampton K, Wrana JL, Attisano L, Szidonya J, Cassill JA, Massague J, Hoffmann FM (1994) Identification of two *Drosophila* Bone Morphogenetic Protein Type 1 receptors in *Drosophila* and evidence that Brk25D is a *decapentaplegic* receptor. Cell 78: 239–250

Phillips DJ, de Kretser DM (1998) Follistatin: a multifunctional regulatory protein. Front Neuroendocrinol 19:287–322

Piccolo S, Agius E, Leyns L, Bhattacharyya S, Grunz H, Bouwmeester T, De Robertis EM (1999) The head inducer Cerberus is a multifunctional antagonist of Nodal, BMP and Wnt signals. Nature 397:707–710

Piek E, Heldin CH, Ten Dijke P (1999) Specificity, diversity, and regulation in TGF-beta superfamily signaling. FASEB J 13:2105–2124

Podos SD, Ferguson EL (1999) Morphogen gradients: new insights from DPP. Trends Genet 15: 396–402

Podos SD, Hanson KK, Wang Y-C, Ferguson EL (2001) The DSmurf ubiquitin-protein ligase restricts BMP signaling spatially and temporally during *Drosophila* embryogenesis. Developmental Cell 1:567–578

Raftery LA, Sutherland DJ (1999) TGF-beta family signal transduction in *Drosophila* development: from Mad to Smads. Dev Biol 210:251–268

Raftery LA, Twombly V, Wharton K, Gelbart WM (1995) Genetic screens to identify elements of the *decapentaplegic* signaling pathway in *Drosophila*. Genetics 139:241–254

Ramirez-Weber FA, Kornberg TB (1999) Cytonemes: cellular processes that project to the principal signaling center in *Drosophila* imaginal discs. Cell 97:599–607

Ray RP, Wharton KA (2001) Context-dependent relationships between the BMPs *gbb* and *dpp* during development of the *Drosophila* wing imaginal disk. Development 128:3913–3925

Ross JJ, Shimmi O, Vilmos P, Petryk A, Kim H, Gaudenz K, Hermanson S, Ekker SC, O'Connor MB, Marsh JL (2001) Twisted gastrulation is a conserved extracellular BMP antagonist. Nature 410:479–483

Ruberte E, Marti T, Nellen D, Affolter M, Basler K (1995) An absolute requirement for both the type II and type I receptors, Punt and Thick veins, for Dpp signaling in vivo. Cell 80:889–897

Rushlow C, Colosimo PF, Lin MC, Xu M, Kirov N (2001) Transcriptional regulation of the *Drosophila* gene *zen* by competing Smad and Brinker inputs. Genes Dev 15:340–351

Ryoo HD, Marty T, Casares F, Affolter M, Mann RS (1999) Regulation of Hox target genes by a DNA bound Homothorax/Hox/Extradenticle complex. Development 126:5137–5148

Sasai Y, Lu B, Steinbeisser H, De Robertis EM (1995) Regulation of neural induction by the Chd and Bmp-4 antagonistic patterning signals in *Xenopus*. Nature 377:757

Savage C, Das P, Finelli A, Sun C, Baird S, Padgett R (1996) The *C. elegans sma-2*, *sma-3* and *sma-4* genes define a novel conserved family of TGF-β pathway components. Proc Natl Acad Sci USA 93:790–794

Schneyer A, Tortoriello D, Sidis Y, Keutmann H, Matsuzaki T, Holmes W (2001) Follistatin-related protein (FSRP): a new member of the follistatin gene family. Mol Cell Endocrinol 180:33–38

Scott IC, Blitz IL, Pappano WN, Maas SA, Cho KW, Greenspan DS (2001) Homologues of Twisted gastrulation are extracellular cofactors in antagonism of BMP signalling. Nature 410:475–478

Sekelsky JJ, Newfeld SJ, Raftery LA, Chartoff EH, Gelbart WM (1995) Genetic characterization and cloning of *Mothers against dpp*, a gene required for *decapentaplegic* function in *Drosophila melanogaster*. Genetics 139:1347–1358

Shimasaki S, Koga M, Esch F, Cooksey K, Mercado M, Koba A, Ueno N, Ying SY, Ling N, Guillemin R (1988) Primary structure of the human follistatin precursor and its genomic organization. Proc Natl Acad Sci USA 85:4218–4222

Shimell MJ, Ferguson EL, Childs SR, O'Connor MB (1991) The *Drosophila* dorsal-ventral patterning gene *tolloid* is related to human bone morphogenetic protein 1. Cell 67:469–481

Sirard C, Kim S, Mirtsos C, Tadich P, Hoodless PA, Itie A, Maxson R, Wrana JL, Mak TW (2000) Targeted disruption in murine cells reveals variable requirement for Smad4 in transforming growth factor beta-related signaling. J Biol Chem 275:2063–2070

Smith WC (1999) TGF beta inhibitors. New and unexpected requirements in vertebrate development. Trends Genet 15:3–5

Spencer FA, Hoffmann FM, Gelbart WM (1982) Decapentaplegic: a gene complex affecting morphogenesis in *Drosophila* melanogaster. Cell 28:451–461

Srinivasan S, Rashka KE, Bier E (2002) Creation of a Sog morphogen gradient in the *Drosophila* embryo. Dev Cell 2:91–101

St Johnston RD, Hoffmann FM, Blackman RK, Segal D, Grimaila R, Padgett RW, Irick HA, Gelbart WM (1990) Molecular organization of the *decapentaplegic* gene in *Drosophila melanogaster*. Genes Dev 4:1114–1127

Staehling-Hampton K, Laughon AS, Hoffmann FM (1995) *Drosophila* protein related to the human zinc finger transcription factor PRDII/MBPI/HIV-EP1 is required for *dpp* signaling. Development 121:3393–3403

Stroschein SL, Wang W, Zhou S, Zhou Q, Luo K (1999) Negative feedback regulation of TGF-beta signaling by the SnoN oncoprotein. Science 286:771–774

Sugino K, Kurosawa N, Nakamura T, Takio K, Shimasaki S, Ling N, Titani K, Sugino H (1993) Molecular heterogeneity of follistatin, an activin-binding protein. Higher affinity of the carboxyl-terminal truncated forms for heparan sulfate proteoglycans on the ovarian granulosa cell. J Biol Chem 268:15579–15587

Takagi T, Harada J, Ishii S (2001) Murine Schnurri-2 is required for positive selection of thymocytes. Nat Immunol 2:1048–1053

Tanimoto H, Itoh S, ten Dijke P, Tabata T (2000) Hedgehog creates a gradient of DPP activity in *Drosophila* wing imaginal discs. Mol Cell 5:59–71

Teleman AA, Cohen SM (2000) Dpp gradient formation in the *Drosophila* wing imaginal disc. Cell 103:971–980

Torres-Vazquez J, Park S, Warrior R, Arora K (2001) The transcription factor Schnurri plays a dual role in mediating Dpp signaling during embryogenesis. Development 128:1657–1670

Torres-Vazquez J, Warrior R, Arora K (2000) *schnurri* is required for *dpp*-dependent patterning of the *Drosophila* wing. Dev Biol 227:388–402

Tsang M, Kim R, de Caestecker MP, Kudoh T, Roberts AB, Dawid IB (2000) Zebrafish nma is involved in TGFbeta family signaling. Genesis 28:47–57

Tsukazaki T, Chiang TA, Davison AF, Attisano L, Wrana JL (1998) SARA, a FYVE domain protein that recruits Smad2 to the TGFbeta receptor. Cell 95:779–791

Tsuneizumi K, Nakayama T, Kamoshida Y, Kornberg TB, Christian JL, Tabata T (1997) *Daughters against dpp* modulates *dpp* organizing activity in *Drosophila* wing development. Nature 389: 627–631

Udagawa Y, Hanai J, Tada K, Grieder NC, Momoeda M, Taketani Y, Affolter M, Kawabata M, Miyazono K (2000) Schnurri interacts with Mad in a Dpp-dependent manner. Genes Cells 5:359–369

Vilmos P, Gaudenz K, Hegedus Z, Marsh JL (2001) The Twisted gastrulation family of proteins, together with the IGFBP and CCN families, comprise the TIC superfamily of cysteine rich secreted factors. Mol Pathol 54:317–323

Waltzer L, Bienz M (1999) A function of CBP as a transcriptional co-activator during Dpp signalling. EMBO J 18:1630–1641

Wang T, Danielson PD, Li B-Y, Shah PC, Kim SD, Donahoe PK (1996) The p21RAS farnesyltransferase α subunit in TGF-β and activin signaling. Science 271:1120–1122

Wang T, Donahoe PK, Zervos AS (1994) Specific interaction of type I receptors of the TGF-beta family with the immunophilin FKBP-12. Science 265:674–676

Wharton KA, Thomsen GH, Gelbart WM (1991) *Drosophila* 60A gene, another transforming growth factor beta family member, is closely related to human bone morphogenetic proteins. Proc Natl Acad Sci USA 88:9214–9218

Wharton KA, Ray RP, Gelbart WM (1993) An activity gradient of *decapentaplegic* is necessary for the specification of dorsal pattern elements in the *Drosophila* embryo. Development 117: 807–822

Wharton KA, Cook JM, Torres-Schumann S, de Castro K, Borod E, Phillips DA (1999) Genetic analysis of the bone morphogenetic protein-related gene, *gbb*, identifies multiple requirements during *Drosophila* development. Genetics 152:629–640

Wisotzkey RG, Mehra A, Sutherland DJ, Dobens LL, Liu X, Dohrmann C, Attisano L, Raftery LA (1998) Medea is a *Drosophila* Smad4 homolog that is differentially required to potentiate DPP responses. Development 125:1433–1445

Wrana JL, Attisano L (2000) The Smad pathway. Cytokine Growth Factor Rev 11:5–13

Xie T, Finelli AL, Padgett RW (1994) The *Drosophila saxophone* gene: a serine-threonine kinase receptor of the TGF-β superfamily. Science 263:1756–1759

Xu X, Yin Z, Hudson JB, Ferguson EL, Frasch M (1998) Smad proteins act in combination with synergistic and antagonistic regulators to target Dpp responses to the *Drosophila* mesoderm. Genes Dev 12:2354–2370

Yao D, Dore JJ Jr, Leof EB (2000) FKBP12 is a negative regulator of transforming growth factor-beta receptor internalization. J Biol Chem 275:13149–13154

Yu K, Srinivasan S, Shimmi O, Biehs B, Rashka KE, Kimelman D, O'Connor MB, Bier E (2000) Processing of the *Drosophila* Sog protein creates a novel BMP inhibitory activity. Development 127:2143–2154

Zhang Y, Chang C, Gehling DJ, Hemmati-Brivanlou A, Derynck R (2001) Regulation of Smad degradation and activity by Smurf2, an E3 ubiquitin ligase. Proc Natl Acad Sci USA 98:974–979

Zheng X, Wang J, Haerry TE, Wu AY, Martin J, O'Connor MB, Lee CH, Lee T (2003) TGF-beta signaling activates steroid hormone receptor expression during neuronal remodeling in the *Drosophila* brain. Cell 112:303–315

Zhu H, Kavsak P, Abdollah S, Wrana JL, Thomsen GH (1999) A SMAD ubiquitin ligase targets the BMP pathway and affects embryonic pattern formation. Nature 400:687–693

The Chemokine Networks in Sponges: Potential Roles in Morphogenesis, Immunity and Stem Cell Formation

W.E.G. Müller, M. Wiens, I.M. Müller, and H.C. Schröder

Abstract. Porifera (sponges) are now well accepted as the phylum which branched off first from the common ancestor of all metazoans, the Urmetazoa. The transition to the Metazoa became possible because during this phase, cell–cell as well as cell–matrix adhesion molecules evolved which allowed the formation of a colonial stage of animals. The next prerequisite for the evolution to the Urmetazoa was the establishment of an effective immune system which, flanked by apoptosis, allowed the formation of a first level of individuation.

In sponges (with the model *Suberites domuncula* and *Geodia cydonium*), the main mediators of the immune responses are the chemokines. Since sponges lack a vascular system and consequently blood cells (in the narrow sense), we have used the term chemokines (in a broad sense) to highlight that the complex network of intercellular mediators initiates besides differentiation processes also cell movement. In the present review, the cDNAs encoding the following chemokines were described and the roles of their deduced proteins during self-self and nonself recognition outlined: the allograft inflammatory factor, the glutathione peroxidase, the endothelial-monocyte-activating polypeptide, the pre-B-cell colony-enhancing factor and the myotrophin as well as an enzyme, the (2–5)A synthetase, which is involved in cytokine response in vertebrates.

A further step required to reach the evolutionary step of the integrated stage of the Urmetazoa was the acquisition of a stem cell system. In this review, first markers for stem cells (mesenchymal stem cell-like protein) as well as for chemokines involved in the maintenance of stem cells (noggin and glia maturation factor) are described at the molecular level, and a first functional analysis is approached. Taken together, it is outlined that the chemokine network was essential for the establishment of metazoans, which evolved approximately 600 to 800 million years ago.

1 Introduction

Since the introduction of molecular biological techniques with the aim to sequence informational genes encoding structural or regulatory proteins, it became evident that the phylum Porifera (sponges) must be grouped together with the other multicellular animals into one kingdom (Müller 1995). The divergence of Porifera from the common ancestor, the Urmetazoa (Müller 2001), occurred approximately 600 to 800 million years ago (MYA) as calculated from the sequence similarities among galectin molecules within the Metazoa (Hirabayashi and Kasai 1993; Pfeifer et al. 1993). Most of the molec-

Institut für Physiologische Chemie, Abteilung Angewandte Molekularbiologie, Universität, Duesbergweg 6, 55099 Mainz, Germany

Progress in Molecular and Subcellular Biology
A. Beschin (Ed.) Invertebrate Cytokines

ular studies on this topic have been performed with the two marine demosponges *Suberites domuncula* and *Geodia cydonium* (our group), and with the freshwater sponge *Ephydatia fluviatilis* (Hoshiyama et al. 1998).

Other pressing questions were for the origin of the metazoan Bauplan or the formation of the polarity in sponges. Also in this context, recent molecular data provided a solid basis for the existence of molecules that regulate developmental axes in demosponges (*G. cydonium*, *S. domuncula* and *E. fluviatilis*). At first, such molecules had to be found which allow tuned cell–cell as well as cell–matrix interactions. The major breakthrough came with the detection of the key metazoan molecules involved in cell adhesion, such as the cell-surface bound integrins (Pancer et al. 1997; reviewed in Müller 1997) and their interacting extracellular molecules with collagen in the first place (Garrone 1998; Krasko et al. 2000).

In addition to those receptors and their interacting ligands which are characteristic for metazoans, sponges have developed a further special set of adhesion molecules, the aggregation factor (AF) and the cell surface-bound aggregation receptor (AR). Two sponges, *Microciona prolifera* and *G. cydonium*, served as models to elucidate their identities (for review Burger and Jumblatt 1977; Müller 1982). In 1973, the two groups succeeded to isolate and purify from *M. prolifera* (Henkart et al. 1973) and *G. cydonium* (Müller and Zahn 1973) an extracellular particle, AF, which promotes the species-specific aggregation of sponge cells. The AFs were characterized as high-molecular weight complexes (Cauldwell et al. 1973; Zahn et al. 1976) which are assembled from a series of proteins that are bound either covalently or noncovalently to the core structure (Müller et al. 1979; Misevic et al. 1982; Wagner-Hülsmann et al. 1996). The AF which mediates cell–cell aggregation depends on the presence of Ca^{2+} (Burger and Jumblatt 1977; Müller 1982) and binds to the cell surface-bound aggregation receptor (AR) (Weinbaum and Burger 1973; Müller et al. 1976). One of the adhesion molecules associated with the AF is the galectin which was isolated and cloned (Pfeifer et al. 1993). In *G. cydonium* the galectin links the AF to the membrane-associated AR (Wagner-Hülsmann et al. 1996). Also, the putative AR was cloned from *G. cydonium* and found to comprise fourteen scavenger receptor cysteine-rich (SRCR) domains, six short consensus repeats (SCR), a C-terminal transmembrane domain and a cytoplasmic tail (Blumbach et al. 1998). Burger's group succeeded to identify in the AF-complex from *M. prolifera*, a proteoglycan-like core structure, from which a protein termed MAFp3 has subsequently been cloned (Fernandez-Busquets et al. 1996; Fernandez-Busquets and Burger 1997). This protein, MAFp3, is likely to be entrapped into a polysaccharide cover (Fernandez-Busquets et al. 1996). Sequence analysis revealed that the gene encoding MAFp3 is highly polymorphic and might be involved in the cell adhesion system during sponge allogeneic reactions (Fernandez-Busquets and Burger 1997). In addition, evidence has been presented indicating that homologous AF complexes might interact with each other (Fernandez-Busquets and Burger 1999). It should be stressed here that besides protein–protein interactions, glycan–glycan binding reactions might contribute to AF-mediated cell–cell adhesion reactions (Misevic et al. 1982). Recently, further molecules of the AF complex of *G. cydonium* as well as of the extracellular matrix have been identified and cloned (Schütze et al. 2001a,b).

Besides the molecules which allow the interaction of cells with the extracellular milieu, those receptors had to be identified in sponges which are primarily involved in the transduction of extracellular signals to the interior of the cells, finally resulting in a modulation of gene transcription. The first gene/protein identified in this respect was the receptor tyrosine kinase from *G. cydonium* which comprises a surprisingly high sequence similarity to the related molecules found in Protostomia as well as Deuterostomia (Müller and Schäcke 1996). In addition, also one seven-pass transmembrane receptor was cloned and functionally studied, the metabotropic glutamate receptor (Perovic et al. 1999). This receptor can be considered as the phylogenetically oldest neuronal receptor of Metazoa.

The construction of the Bauplan of the phylum Porifera is controlled by a series of transcription factors with the homeobox genes in the first instance. Besides more sponge-specific homeobox genes, e.g. those belonging to the *POU-*, *Cad-*, *En-*, *Msx-* or *NK-2* family (Seimiya et al. 1994; Richelle-Maurer and van de Vyver 1999; Manuel and Le Parco 2000), molecules with high similarities to those of other metazoan phyla that are involved in Bauplan formation, e.g. Pax-2/5/8 class (Hoshiyama et al. 1998) or the Lim-class homeodomain proteins (Wiens et al. 2003), have been discovered. Furthermore, we have identified in *S. domuncula* the expression of *Forkhead/T-box* genes (Adell et al. 2002) and of a gene encoding the retinoic acid receptor (unpubl.); both classes of transcription factors are known to be involved in morphogenesis of Metazoa. However, like the Cnidaria, the sponges apparently lack the gene clusters allowing a vectorial patterning mechanism.

Porifera have two layers of epithelial cells (pinacocytes), the external epithelium (exopinacocytes) and the internal epithelium (endopinacocytes); the latter one represents the surface of the aquiferous canal system (Simpson 1984). Within these layers which are composed of cells firmly connected through simple cell junctions, a bulky mesohyl compartment exists. In the mesohyl, the somatic cells, predominantly the contractile cells (myocytes) and the skeletal cells (collencytes, sclerocytes), as well as the totipotent archaeocytes exist; all are only loosely attached to each other. The extensive extracellular matrix is composed of collagen, lectin and glycoconjugates (Diehl-Seifert et al. 1985); the latter component is not yet chemically characterized and no cDNA has been cloned. Hence sponges, like the other diploblastic taxon – the Cnidaria, are lacking a circulatory system. A defined muscle system is also absent in sponges.

It is generally agreed that archaeocytes are totipotent cells from which the somatic cells, e.g. myocytes, collencytes/sclerocytes or pinacocytes, derive. In addition, the archaeocytes give rise to the likewise totipotent germ cells (Simpson 1984; Fig. 1). Hence, these cells (archaeocytes) should be considered

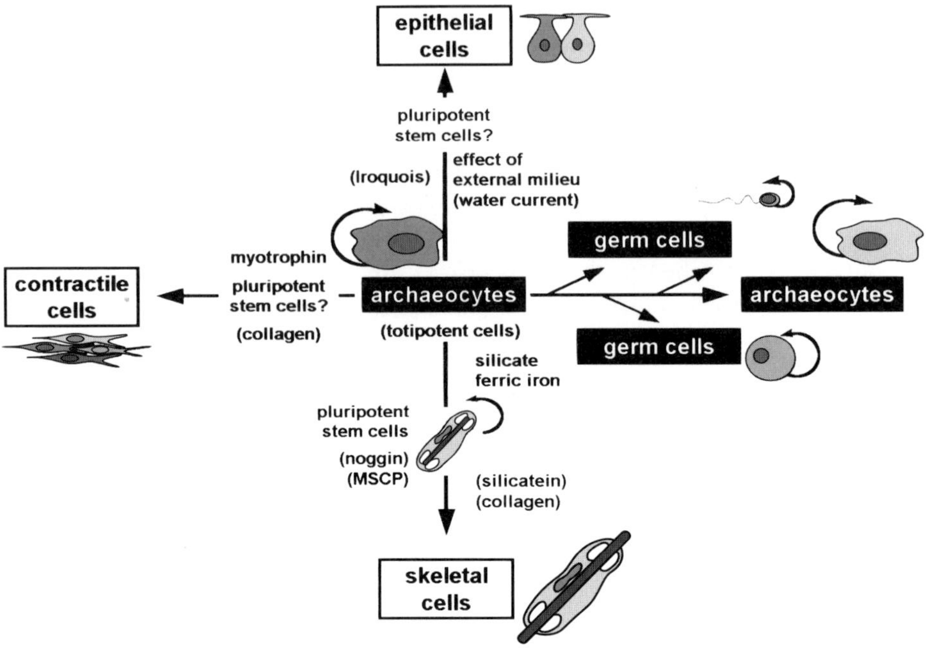

Fig. 1. Schematic outline of the postulated development of the totipotent archaeocytes to the germ cells on one side and to the three major differentiated cells, the epithelial, the contractile and the skeletal cells. It is postulated that the latter differentiation processes occur through precursor pluripotent stem cells. The major factors triggering these processes as well as the genes which are upregulated during these events are indicated

as stem cells since they comprise the function of self-renewal and the plasticity to differentiate into a distinct path. By this property, the archaeocytes/stem cells have the function to repair damage and also the capacity to extend organismal life beyond that of the component cells (Spradling et al. 2001). Two mechanisms might control the generation of mature cells from stem cells, either an internal program in the stem cells resulting in the formation of unequal division under formation of daughter cells of different fate, or, external factors emanating from neighboring differentiated cells. More experimental evidence supports the latter view (Spradling et al. 2001). It can be postulated that with the transition to the multicellular organisms (here we focus on animals) a "compartmentalization" of cells into totipotent cells on one side and more differentiated cells with limited differentiating potency on the other side must have been established.

A further elucidation of the differentiation potential of the stem cells and the potential signals or other intercellular interactions which control their differentiation pattern can be obtained by comparative studies using the most original metazoan organisms, the sponges. One clue for reliable answers should be seen in the identification and characterization of genes coding for those molecules which (1) are characteristic for the identification of stem cells and (2) limit – either stimulate differentiation or determine the totipotent property – the development of stem cells. If these molecules can be identified in sponges, then these organisms are a basic model system useful for the understanding of the stem cell concept in Metazoa. In this review we give first evidence of such polypeptides in sponges.

Considering that no vessel system is present in sponges, the finding that these animals are provided with a powerful immune system came very surprising (for review Müller et al. 1999b). Applying the parabiotic attachment technique or the insertion technique for allo/autografting experiments, it could be demonstrated that sponges, with the model systems of *S. domuncula* and *G. cydonium*, can effectively distinguish between self and nonself. This property of the Porifera was essential for their success to survive extinction during the period from the Proterozoic until present. Since sponges are sessile filter feeders – some of which have the capacity to process their own volume of water every 5 s in order to extract nutritious material (Vogel 1977) – it can be assumed that they are exposed to a huge amount of bacteria and/or viruses present in the seawater (Gonzales and Moran 1997). To cope with these threats, sponges have developed an efficient chemical defense system (Sarma et al. 1993; Proksch 1994) as well as humoral and cellular defense mechanisms (Müller et al. 1999b).

Even though sponges comprise – at first glance – a loosely composed cellular organization, the sponge individual is a stable cell population and, at no stage in their biological cycle, the cells are dispersed (Garrone 1978). Consequently, they must have acquired also a complex network of intercellular mediators, of cytokines and chemokines. In vertebrates, both classes of signaling molecules are members of the cytokine superfamily; the cytokines are produced primarily by lymphocytes and mediate immune responses, while chemokines are produced both by phagocytes and other tissue cells (endothelium, fibroblasts) and usually contain two internal disulfide loops. In order to avoid any conflicting view, known from previous studies with invertebrate cytokines, in the following review only those polypeptides are summarized which have been cloned and found to comprise high sequence similarity to molecules known from triploblasts. Since functional studies suggest that these polypeptides are involved in the control of proliferation as well as of differentiation processes and are thought to mediate cell movement, we *operationally term these factors* ***chemokines*** *in the broadest sense.*

2
The Model System: *Suberites domuncula* (Demospongiae)

This sponge was selected for our work since it turned out that both the animals and the cells can be cultivated in aquaria/in vitro cultures in a convenient manner (Le Pennec et al. 2003). Based on this advantage, studies on the immune reactions are routinely possible.

2.1 Transplantation Studies

It has been well established at least since 1842 (Laurent 1842) for freshwater sponges and since 1886 for marine sponges (Buccich 1886), e.g. *Euspongia officinalis*, that sponges are characterized by unusual regeneration potencies. This property has been successfully used to farm sponges in large mariculture projects (Moore 1908). Scientifically, the remarkable regeneration capabilities have been studied by Wilson (1907), Galtsoff (1925) and Humphreys (1963) in classical re-aggregation studies. Smith and Hildemann in their extensive review (Smith and Hildemann 1986) have grouped sponge allo-immune responses in experimental transplantations into two major rejection processes. Some species may form barriers to separate from nonself tissue; e.g. the marine sponge *Axinella verrucosa* (Buscema and van de Vyver 1983) or the freshwater sponge *Ephydatia muelleri* (Mukai 1992), while others may react by cytotoxic factors which destroy the transplant; e.g. the marine sponges *Callyspongia diffusa* (Hildemann et al. 1979) or *Geodia cydonium* (Pfeifer et al. 1992).

In our studies (Pancer et al. 1996), we used two different techniques for allo- and autografting experiments: (1) the insertion technique and (2) the parabiotic attachment. While autografts always fused, allografts fused only in very rare cases (less than 5%). In addition we could show that, depending upon the method of grafting, either a barrier-like structure is formed or a "cytotoxic" response to nonself tissue is induced. The techniques we used have been described in detail (Müller et al. 1999b). In brief, for the insertion technique (primarily used with *G. cydonium*; Fig. 2A-a), tissue pieces removed with a cork drill from one specimen were inserted into holes of the recipients (Fig. 2A-b) which had a slightly narrower diameter. All autografts fused and no boundary line was finally seen (Fig. 2A-d, left: *au*). Allografts initially fused; however, after approximately 2 to 3 days the rejected graft tissue formed a pronounced demarcation boundary and underwent apoptotic degeneration (Fig. 2A-c and A-d, *al*) and finally resorption.

For the parabiotic attachment technique, slices approximately 1–2 cm thick, were cut from the surface of animals (here *S. domuncula* was used; Fig. 2B-a). Pieces matching in size were attached to each other with their cut side and loosely fixed by nylon fibers (experiment with *S. domuncula*; Fig. 2B-b). One day later, the fibers could be removed, as the grafts/parabionts already adhered to each other. Since *S. domuncula* occurs in nature in red, orange, whitish, blue or as a mixture of these colors (Fig. 2B-a), it is already possible without any staining techniques to distinguish the transplants. Autografts from *S. domuncula* started to fuse approximately 2 days after grafting; after 6 days the zone where the two grafts were attached was no longer visible, both by eye inspection (Fig. 2B-d) and by microscope analysis after sectioning and staining. In contrast, allografts started to reject each other after an initial period of fusion (duration 2 to 3 days); a gap became visible ~4 days after grafting (Fig. 2B-c).

2.2 The Primmorphs

It was a painful period until we realized that – according to our experience – it is not yet possible to cultivate sponge cells in vitro in a single cell state. During our work focusing on cell-surface receptors (Müller 1997), it became evident that a pronounced division of sponge cells can be obtained only if the cells remain in close cellular contact (Custodio et al. 1998; Müller et al. 1999a).

A detailed review of the initiation and maintenance of the primmorph culture was given recently (Le Pennec et al. 2003). In brief, dissociated cells (Fig. 2C-a) are transferred into the Ca^{2+} and Mg^{2+}-containing seawater/or natural seawater enriched with 1 to 10% Dulbecco's MEM (with glutamine and without $NaHCO_3$). Usually after 5 days primmorphs are formed from dissociated cells (Fig. 2C-d). The primmorphs are characterized by the presence of proliferating cells as well as by a characteristic histology. Microscopic analysis of cross sections through primmorphs showed an almost complete single-cellular layer of epithelial-like cells surrounding the primmorphs (Custodio et al. 1998; Müller et al. 1999a). The cells in the squamous epithelium of the primmorphs are pinacocytes, as judged from their flattened, fusiform extensions and their prominent nucleus. The cells inside the primmorphs are primarily archaeocytes/spherulous cells. In order to ascertain that the cells which are reorganized into primmorphs indeed originate from the species *S. domuncula*, primmorphs were cross-sectioned and reacted with antiserum raised against *S. domuncula* cells. The immunofluorescence analysis of these cross sections showed that almost all cells in the primmorphs were stained brightly with the antiserum (Grebenjuk et al. 2002). Control sections, incubated with preimmune serum did not show any reactions.

In some primmorph cultures, the basic culture medium, seawater with Dulbecco's MEM, was supplemented with inorganic components, silicon and Fe^{3+}, to optimize the growth of the primmorphs (Krasko et al. 2002).

2.3 The Mixed Sponge Cell Reaction (MSCR) Assay

Recently, an assay system has been established which allows the determination of immune reaction in vitro (Müller et al. 2002). In the mixed sponge cell reaction (MSCR) assay, dissociated cells (Fig. 2C-a) from the same individual or from different individuals are mixed at equal cell concentrations. If cells from different individuals were maintained separately for 12 h in seawater, they

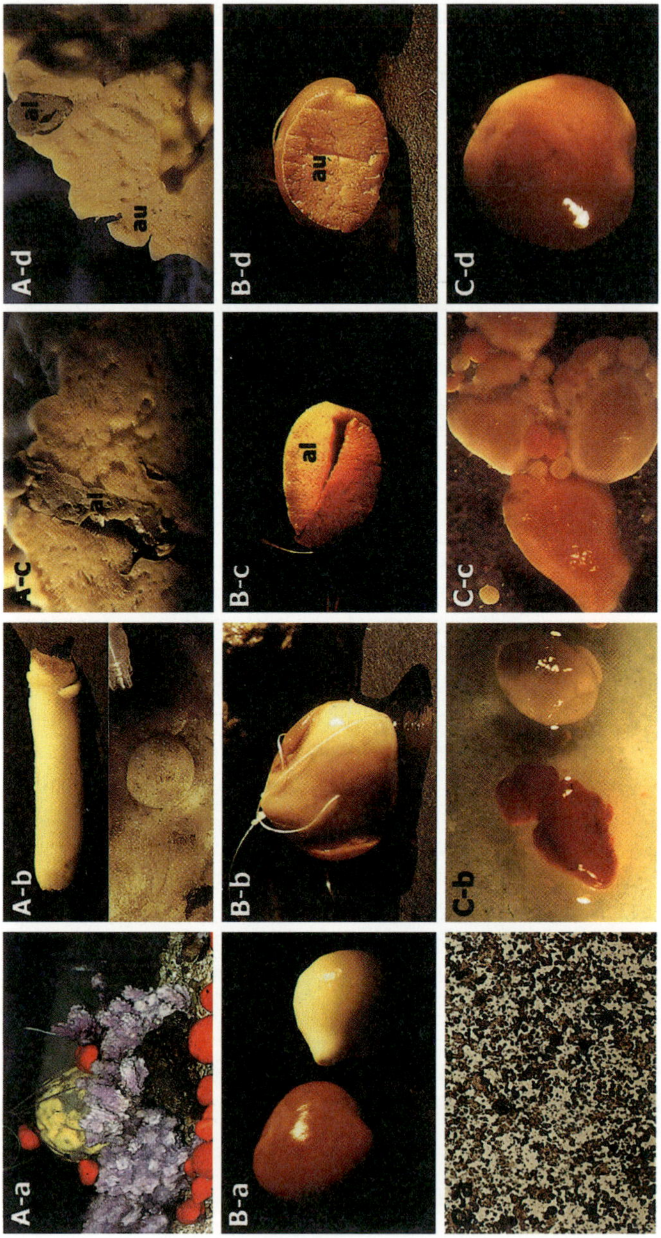
A-a
A-b
A-c
al
A-d
au
al
B-a
B-b
B-c
al
B-d
au
C-b
C-c
C-d

formed small aggregates; then, the aggregates either from the same (autogeneic MSCR) or from different individuals (allogeneic MSCR) were mixed and transferred to natural seawater/DMEM medium. In order to distinguish by eye inspection the aggregation pattern of cells in an allogeneic MSCR, the color of one donor (and hence of the isolated cells) was white, while the second one was red. For the experiments in autogeneic MSCR, the color of the donor was either red or white. Two days after mixing the aggregates, they were inspected. After a further period of five days, primmorphs had been formed from the aggregates.

If cells from the same individual were mixed, autogeneic MSCR, 2 to 8 mm large primmorphs were formed during the incubation period (Fig. 2C-d). For the analysis of aggregation/primmorph formation from cells derived from different individuals, in the allogeneic MSCR, cells from white- and red-colored animals were used. In these assays, the aggregates separated after two days, indicating that during the allogeneic MSCR the cells form individual-specific aggregates. If these aggregates were incubated for additional 5 days, white- or red-colored primmorphs were formed (Fig. 2C-b). In one series of experiments, cells from different red specimens were used for an allogeneic MSCR; also here a strong separation of the primmorph occurred after their formation (Fig. 2C-c). These findings confirm an earlier study in which it was reported that aggregates from different individuals form, after a short transient contact period, different (distinct) and individual aggregates (Humphreys 1994).

◀

Fig. 2. Tissue/cell recognition in sponges. **A** and **B** Transplantation studies. **A** Recognition of autografts and allografts of the sponge *G. cydonium* using the insertion technique for grafting. (**A–a**) The sponge *G. cydonium* (*yellow*) kept in the aquarium for more than 2 years which is surrounded by *S. domuncula* (*red*) and *Dysidea avara* (*violet*); ×0.05. **A–b** Tissue pieces were removed with a cork drill from one *G. cydonium* specimen and inserted into holes in the recipients, which had a slightly narrower diameter; ×0.6. **A–c** Allografts initially fused together but underwent already apoptotic degeneration after 3 days a process which ultimately resulted in resorption; ×0.5. **A–d** Autograft (*au*) fused after 5 days; into the same specimen an allograft was inserted (*al*) which was rejected; ×0.5. **B** Autograft fusion and allograft rejection of transplants from *S. domuncula*. **B–a** Two specimens in red and white are shown; ×0.5. **B–b** Two grafts are tied together with a fiber; ×1. **B–c** 4-day-old allograft (cross section) from a red and a whitish specimen which reject each other; ×1. **B–d** An autograft 4 days (cross section) after transplantation; ×1. **C** Cell recognition in sponges/formation of primmorphs from cells of *S. domuncula*. **C–a** Dissociated single cell suspension and small cell clumps; ×100. The technique of mixed sponge cell reaction (MSCR) was applied to study the immune response of sponges in vitro. Small aggregates from the same individual (autogeneic MSCR [mixed sponge cell reaction]) or from different individuals (allogeneic MSCR) were placed together and cultured for 2 days. In the allogeneic MSCR the aggregates sort out and form individual clumps, as can be seen from the different colors of the aggregates/primmorphs, originating from white and red individuals; **C-b** after additional 5 days primmorphs are formed; ×10. **C–c** In a parallel experiment, the allogeneic MSCR was performed with cells from two red specimens; ×20. **C-d** In the autogeneic MSCR the aggregates formed 1-2 mm large clumps and 2 days later primmorphs; ×2.5

3 Chemokines: Effects During Immune Response

Our transplantation studies have revealed a series of chemokine molecules which are expressed in the contact zone between the host and the graft. Among those are the following molecules, which share the highest similarity to triploblastic animals.

3.1 Allograft Inflammatory Factor

The most detailed studied chemokine in the sponge system is the allograft inflammatory factor (AIF) (Kruse et al. 1999; Müller et al. 2002). The expression of this factor is known to be highly upregulated in cytokine-activated macrophages in rats (Utans et al. 1995). Moreover, AIF-1 is expressed in rejecting allografts (Utans et al. 1995) and found to be involved in inflammatory response associated with human cardiac transplant rejection (Utans et al. 1996). In addition, it was found that the expression of AIF-1 is upregulated in the pathogenesis of some hormone-dependent diseases (Chen et al. 1997).

3.1.1 Sequence

The *S. domuncula* allograft inflammatory factor-related sequence was deduced from a cDNA which was obtained by PCR technique applying primers designed against the characteristic EF-hand-like motif (Utans et al. 1995; Imai et al. 1996) as described (Kruse et al. 1999). The cDNA *SDAIF* has a size of 523 nucleotides (accession number Y18439). The deduced 144 amino acid-long sequence of the complete cDNA termed AIF_SUBDO has a putative size (M_r) of 16,637 and an isoelectric point (pI) of 4.96. Northern blot analysis performed with the sponge *SDAIF* clone as a probe yielded one prominent band of approximately 0.8 kb, confirming that a full length cDNA was isolated.

In the sponge AIF-1-related protein the EF-hand-like motif is found between amino acid 55 to amino acid 67 (Fig. 3A). This domain precedes the internal 39-residues long motif characteristic of peptide hormones; it is framed by the KR–GK amino acids which are putative cleavage sites for prohormone convertases and peptide amidation enzymes (Chen et al. 1997). Within the potential peptide hormone, the typical KK motif is found. Immediately adjacent, one nuclear targeting signal sequence is located (Fig. 3A).

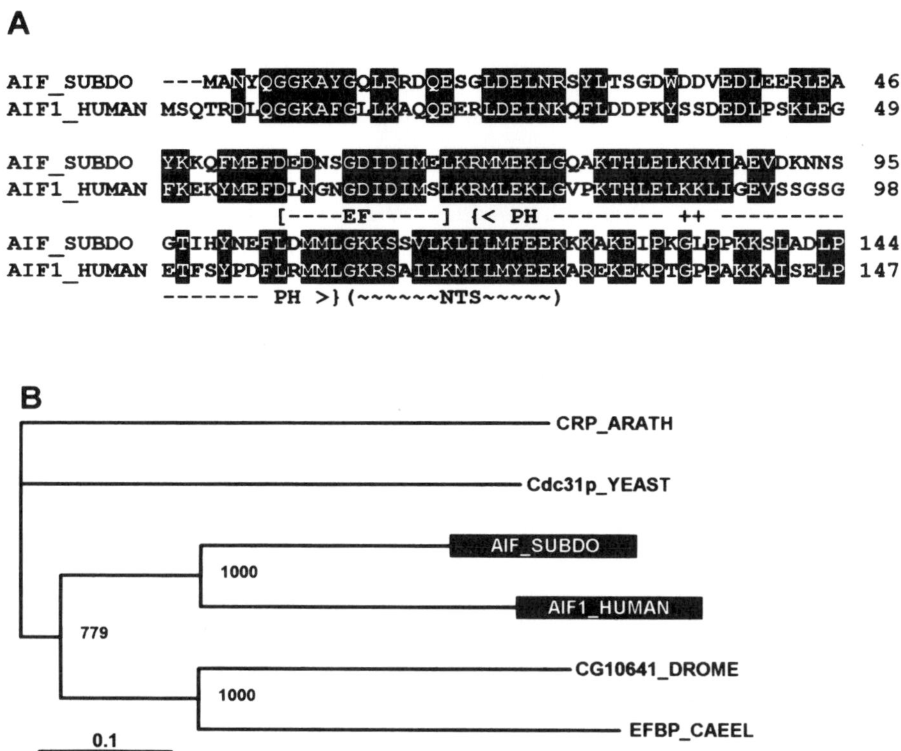

Fig. 3. *S. domuncula* AIF-1-related molecule. **A** Alignment of the deduced protein, AIF_SUBDO with the related sequence from human, the allograft inflammatory factor 1, isoform 3 (AIF1_HUMAN, accession number NP_001614; Iris et al. 1993). Within the sponge AIF-1-related protein the location of the EF-hand-like motif ([–EF–]), the sequence motif characteristic for posttranslational processing of peptide hormones ({<PH>}) with the internal KK motif (++) as well as the potential nuclear targeting sequence (~NTS~) are indicated. Residues conserved in the two sequences are shown in inverted type. **B** Rooted phylogenetic tree constructed with the *S. domuncula* AIF-1-related molecule [AIF_SUBDO], the CG10641 gene product from *Drosophila melanogaster* (CG10641_DROME, NP_609925), the EF hand calcium-binding protein from *Caenorhabditis elegans* (EFBP_CAEEL, NP_496961), the calcium-binding protein component of spindle pole bodies (Cdc31p) from *Saccharomyces cerevisiae* (Cdc31p_YEAST, NP_014900) and the calmodulin-related protein from *Arabidopsis thaliana* (CRP_ARATH, NP_014900). The analysis was performed by neighbour-joining; the numbers at the nodes are an indication of the level of confidence (given in percentage) for the branches as determined by bootstrap analysis (1000 bootstrap replicates). The *scale bar* indicates an evolutionary distance of 0.1 aa substitutions per position in the sequence. The distantly related aa sequence from the plant was used as outgroup

3.1.2
Phylogenetic Relationship

In this analysis and using the databases (Advanced Blast [http://www.ncbi.nlm.nih.gov/blast/blast.cgi]), those sequences from one deuterostomian (human), two protostomians – one insect (*Drosophila melanogaster*) and one nematode (*Caenorhabditis elegans*) – from the yeast *Saccharomyces cerevisiae* and from the plant *Arabidopsis thaliana* were selected which shared highest sequence similarity to the sponge sequence. A phylogenetic tree was constructed and rooted with the plant sequence. This approach reveals that the sponge AIF-1-related protein shares a high similarity (51% identical amino acids / 70% similar amino acids) to the human allograft inflammatory factor 1, isoform 3 (accession number NP_001614). More distantly related are the sequences from *D. melanogaster* [the CG10641 gene product] (NP_609925; 20% identical amino acids / 34% similar amino acids) and *C. elegans* [EF hand calcium binding protein] (NP_496961; 10% identical amino acids / 5% similar amino acids). Only distantly related are the molecules from the yeast *S. cerevisiae* [Cdc31p] (NP_014900; 27% identical amino acids / 19% similar amino acids) and the calmodulin-related protein from *A. thaliana* (NP_014900; 19% identical amino acids/9% similar amino acids) (Fig. 3B).

As in most analyses of this kind, the sponge sequence displays a higher sequence similarity to the corresponding human sequence than to the two protostomian sequences from *D. melanogaster* and *C. elegans* (Gamulin et al. 2000). The degrees of similarities can be deduced from the tree; the sponge AIF-1-related protein groups together with the human AIF and is clearly separated from the two invertebrate sequences.

3.1.3
Potential Function

The sponge AIF was the first cytokine/chemokine identified and functionally analyzed in sponges (Kruse et al. 1999; Müller et al. 2002). Especially basing on the technique of differential display of mRNA it became impressively overt, that AIF is one of the major chemokines which is strongly expressed after immune responses of *S. domuncula* with allogeneic tissue. In the first study, tissue slices from the fusion zones were taken, approximately 1.5 mm thick, and used for the extraction of RNA (Kruse et al. 1999). These samples were subsequently used in Northern blot experiments to determine semiquantitatively the expression of *AIF*. The hybridization signals were detected by the "DIG"-labeling procedure and visualized by the chemiluminescence technique using CDP-Star (Roche). Finally, the signals of the Northern blots were scanned with the GS-525 Molecular Imager (Bio-Rad). The relative values for the expression are given as a multifold of the intensities seen in the controls (expression of the gene at time zero).

The graphic representation of the values are depicted in Fig. 4A-1. It is shown that in autografts (self-reaction), no significant change of expression is seen during the 6 days following transplantation. However, if the level of transcripts of *AIF* is determined in the attachment zone between allografts, a strong upregulation of this gene is observed (nonself reaction). Already 2 days after grafting, the steady-state level increased by 2.5-fold and rose further at day 4 (15.5-fold) and day 6 (21.5-fold) (Kruse et al. 1999). Based on this strong alteration of the expression of *AIF* at the contact zone between self and nonself tissue this gene served as the first marker gene to determine if also in the in vitro system, in the MSCR assay, an immune response could be detected on molecular level (Müller et al. 2002). As summarized in Fig. 4A-2, only a slight increase in *AIF* expression is seen during autogeneic MSCR, while on the contrary a strong increase is seen during allogeneic MSCR. At day 2, the level of expression amounted to 7.4-fold if correlated with the level seen in the controls (set to onefold).

From these data we conclude that in the *S. domuncula* graft as well as MSCR system, the chemokine, AIF_SUBDO, is a marker indicative for an ongoing immune response after contact to allogeneic tissue/cells. In addition, it can be assumed that this chemokine might function in *S. domuncula* during processes which follow inflammatory response during transplant rejection as known for mammalian systems (Utans et al. 1996).

3.2 Glutathione Peroxidase

In the same series of transplantation studies with the *S. domuncula* system (Kruse et al. 1999), a further potential chemokine was identified, the glutathione peroxidase (GPX). In early phases of inflammation, occurring during graft recognition (Shiraishi et al. 1997) or during wound healing (Kheradmand et al. 1998) in mammals, reactive oxygen species (ROS) are formed. One of the major enzymes involved in the detoxification of ROS during these processes is GPX (Cikryt et al. 1982). GPX enzymes use glutathione to reduce hydrogen peroxides and other ROS (Cikryt et al. 1982). Recently, one novel non-selenium GPX, also termed antioxidant protein 2 (AOP2; EC 1.11.1.7), was identified and cloned from mice, which was found to be upregulated during the early inflammatory phase of wound healing (Munz et al. 1997).

3.2.1 Sequence

The putative *S. domuncula* glutathione peroxidase was cloned. The 837 nucleotide-long insert of the cDNA, *SDGPX*, has an ORF, GPX_SUBDO, of 217 amino acids (Fig. 5A) and corresponds to a total mass of 24,131 Da. Expression of the *SDGPX* gene was examined by Northern blot analysis.

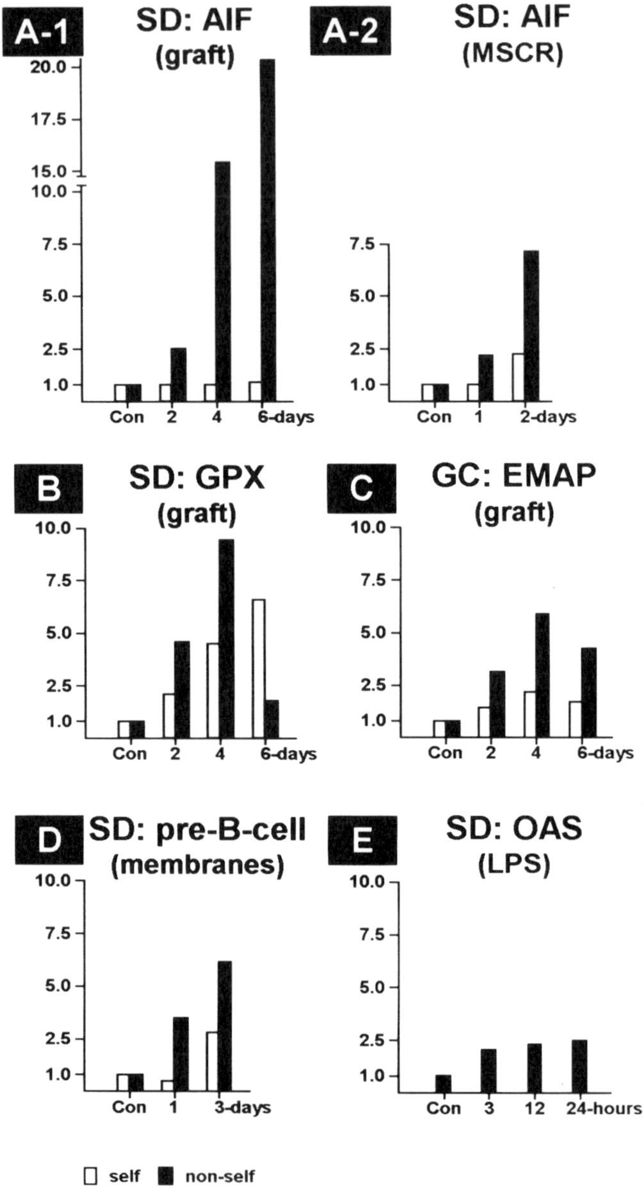

Fig. 4. Expression of chemokines in sponge after contact with self (*open bars*) or nonself tissue/molecules (*closed bars*). The respective cDNAs ending the chemokines were isolated and used in Northern blot analyses for the determination of the steady-state level of expression. In one series of experiments the activity of the (2–5)A synthetase was determined in primmorphs. **A-1** The expression of the chemokine *AIF* (probe: *SDAIF*) was determined in autografts (*open bars*) as well as in allografts (*closed bars*) up to 6 days after grafting. RNA was extracted from the fusion zones and subjected to Northern blot analysis. The relative expression values (correlated with the ones seen in the controls at day zero [Con]) are given; the ordinate gives the level of

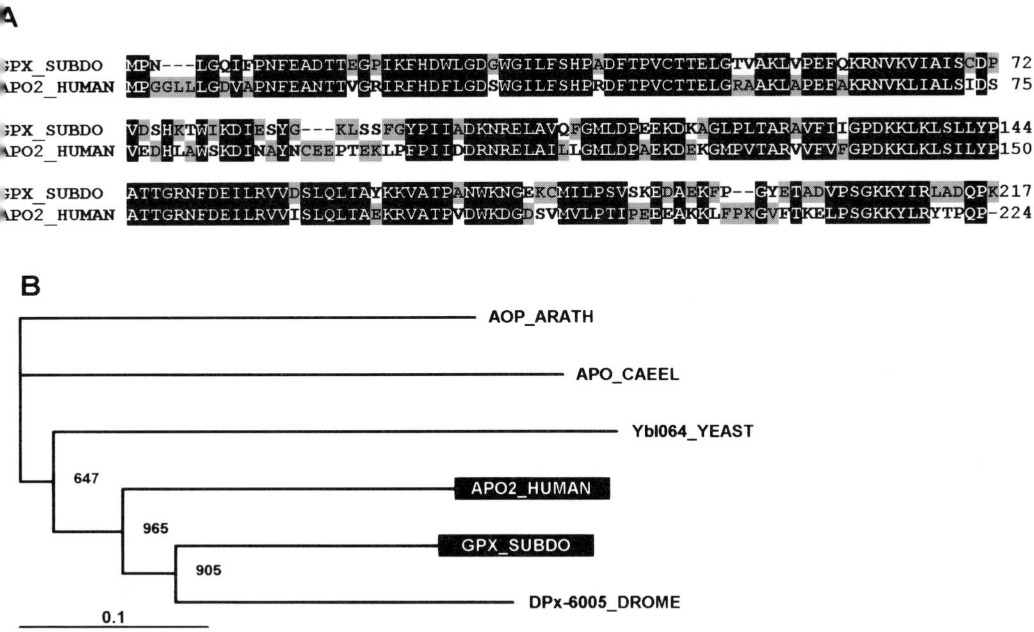

Fig. 5. Sponge GPX-related polypeptide. **A** The protein sequence, from *S. domuncula* (GPX_SUBDO, Y18438), deduced from the cDNA *SDGPX*, is aligned with the corresponding human anti-oxidant protein 2 (non-selenium glutathione peroxidase) (APO2_HUMAN, NP_004896.1; Kim et al. 1997). **B** Rooted phylogenetic tree was constructed with these two sequences and the two protostomian enzymes from *D. melanogaster*, the 1-cys peroxiredoxin DPx-6005 (DPx-6005_DROME, AAG47822; Radyuk et al. 2001) and the *C. elegans* peroxiredoxin (APO_CAEEL, NP_741287). The yeast (Ybl064cp_YEAST, NP_009489) as well as the plant (*A. thaliana* peroxiredoxin (AOP_ARATH, NP_175247; outgroup) enzymes are included as well

◄

expression as multifold of the controls. In a second series **A–2**, the in vitro technique (MSCR) was applied to determine the expression level. During MSCR, primarily during the allogeneic MSCR (*closed bars*) the expression increased in contrast to the one seen in autogeneic MSCR (*open bars*). **B** Expression of *GPX* gene as measured by Northern blot technique and using *SDGPX* as a probe. Again, the level of gene expression was measured in autografts (*open bars*) and allografts (*closed bars*). **C** In a similar series the expression of the gene encoding the endothelial-monocyte-activating polypeptide (EMAP) was monitored in grafts from *G. cydonium* using *GCEMAP2* as a probe. **D** Transcript level of the pre-B-cell CEF in response to autogeneic (*open bars*) or xenogeneic cell membranes from *G. cydonium* (*closed bars*); the sponge *SDPBEF* gene was used as a probe. **E** In this series of experiments the activity of the (2–5)A synthetases was determined in primmorphs from *S. domuncula*. The primmorphs were incubated with 1 μg/ml of LPS for a period of 0 to 24 h. Subsequently crude extracts were prepared and reacted in the enzyme assay (Grebenjuk et al. 2002). Controls (*Con*)

3.2.2
Phylogenetic Relationship

The sponge GPX-related polypeptide shares highest sequence identity/similarity with the human anti-oxidant protein 2, non-selenium glutathione peroxidase (Kim et al. 1997) – 62% identical amino acids and 76% similar amino acids (Fig. 5A) – and with the *D. melanogaster* peroxiredoxin (Radyuk et al. 2001) – 65% identical amino acids and 78% similar amino acids. The other enzymes, including the *C. elegans* enzyme peroxiredoxin, and the yeast as well as the *A. thaliana* peroxiredoxin are only distantly related (Fig. 5B).

3.2.3
Potential Function

Again, Northern blot studies were performed to investigate the expression level of the *S. domuncula* GPX during graft response. It is seen that, especially during the first four days during allograft rejection, a strong upregulation of *GPX* occurs (Fig. 4B). However, it is notable that a considerable increase is also seen in autografts, suggesting that during wound healing this enzyme might be involved in the detoxification of ROS. Hence, we suggest that the expression of the gene encoding the GPX-like enzyme is primarily upregulated during inflammation, irrespectively if caused during the process of autograft fusion or allograft rejection.

3.3
Endothelial-Monocyte-Activating Polypeptide

The endothelial-monocyte-activating polypeptide (EMAP) is considered to be in human an inflammatory cytokine which is released from mononuclear phagocytes during apoptotic processes (Shalak et al. 2001). In the screening study for potential morphogens and/or immune molecules, the EMAP-related molecule was isolated from *G. cydonium* (Pahler et al. 1998).

We have selected the EMAP type II because it causes cell activation, e.g. increase in cytosolic free Ca^{2+} concentration, release of tissue factors, alteration of cell migration and expression of adhesion molecules in endothelial cells as well as in monocytes and granulocytes from human and mouse (Kao et al. 1992, 1994a,b), and it causes angiogenesis (Yoshida et al. 1997).

3.3.1 *Sequence*

The cDNA of the *G. cydonium* EMAP has a size of 592 nucleotides and is complete, as confirmed by Northern blot studies (Pahler et al. 1998). The open reading frame of *GCEMAP2* codes for a 149 amino acid-long protein termed EMAP2_GEOCY, and reveals highest similarity to the human EMAP type II (Kao et al. 1994a,b). As shown in Fig. 6A, this sponge protein matches only the C-terminal part of the human EMAP molecule. From the studies of Kao et al. (1994a), it has been established that the mammalian EMAP molecules are processed and that only the COOH-terminal segment is released from the cells.

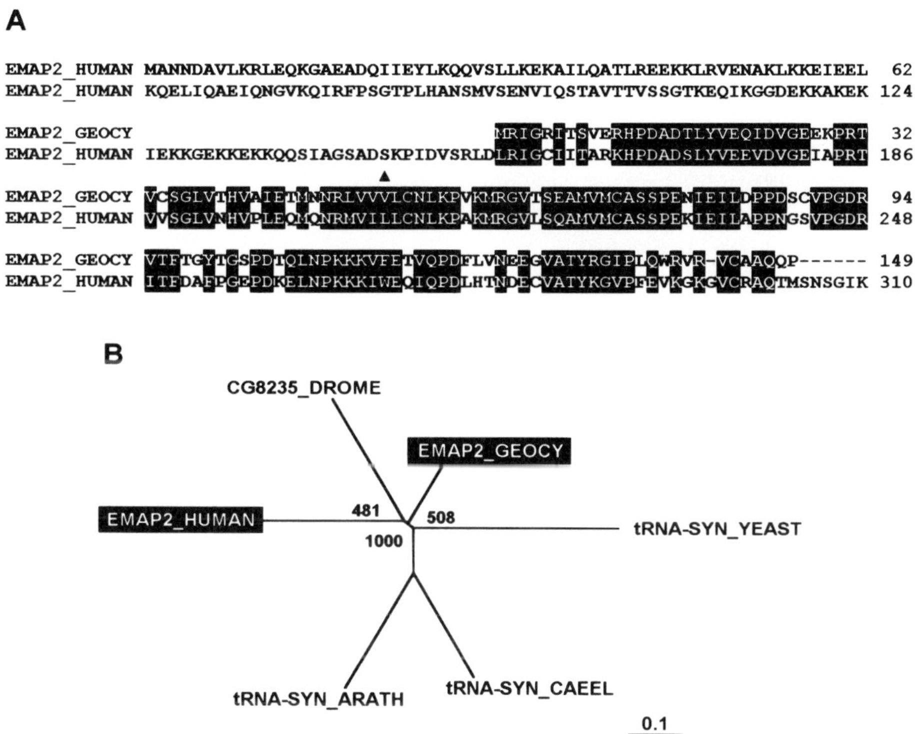

Fig. 6. The *G. cydonium* endothelial-monocyte-activating polypeptide (EMAP). **A** The protein deduced from the complete cDNA was termed EMAP2_GEOCY (Y14947) and is aligned with the human endothelial monocyte-activating protein II precursor (EMAP2_HUMAN, B55053; Kao et al. 1994a, b). The arrow head marks the position of the mature NH_2-terminus in the human EMAP sequence. **B** In order to construct an unrooted phylogenetic tree, the two sequences from *G. cydonium* and human were compared with the related CG8235-PA protein from *D. melanogaster* (CG8235_DROME, NP_610426) as well as the methionyl tRNA synthetases from *C. elegans* (tRNA-SYN_CAEEL, NP_502196), *S. cerevisiae* (tRNA-SYN_YEAST, NP_011410) and *A. thaliana* (tRNA-SYN_ARATH, NP_193114)

This processed form of the human sequence has a size of 165 amino acids and a molecular mass of ~18 kDa (Kao et al. 1994a), and hence is similar to that estimated for the putative EMAP polypeptide (M_r 16,499). A hydropathicity analysis was performed with EMAP2_GEOCY and compared with the one from the mouse EMAP II (Pahler et al. 1998). This approach showed that the distribution of hydropathicity along the sponge sequence is almost identical to that seen in the processed form of the murine sequence, spanning the amino acid residues 151 to 266.

3.3.2
Phylogenetic Relationship

Again, the sponge molecule is closest related to its human counterpart EMAP, the protein II precursor (Kao et al. 1994a,b), with an amino acid identity of 28% and a similarity of 37%. The *D. melanogaster* sequence (CG8235-PA) comprises a relationship of 28% identity and 32% similarity. The methionyl tRNA synthetases from *C. elegans* (9% identity/11% similarity), *S. cerevisiae* (18% identity/24% similarity) and *A. thaliana* (9% identity/12% similarity) show a lower similarity to the sponge sequence. Recently, it was established that the EMAPs from human derive from the aminoacyl-tRNA synthetase complex (Shalak et al. 2001) (Fig. 6B).

3.3.3
Potential Function

The determination of the expression of the sponge *EMAP* gene during graft rejection revealed again that this process of allograft recognition is an active process. In response to contact to allogeneic tissue the expression of this chemokine increases, while the level in autogeneic tissue changes only to a smaller extent (Fig. 4C). In view of the results seen in mammalian systems (Kao et al. 1992, 1994a,b), we adopt the assumption that EMAP is involved in the modulation of cell migration during remodeling processes following transplantation.

3.4
Pre-B-Cell Colony-Enhancing Factor

In a first attempt to elucidate if sponge cells, when cultivated in vitro, can distinguish between self and nonself, primmorphs from *S. domuncula* were incubated with membranes from another demosponge, *G. cydonium* (Müller et al. 1999b). During this process, one chemokine-related molecule could be identified which was found to be upregulated after contact with xenogeneic material, the pre-B-cell colony-enhancing factor (Müller et al. 1999b).

The pre-B-cell colony-enhancing factor (pre-B-cell CEF) is in human (Samal et al. 1994) an effector molecule, a cytokine-related molecule, which is involved in the maturation of B-cell precursors (Ognjanovic et al. 2001). The human pre-B-cell CEF was found to synergize with the interleukin-7 and the stem cell factor (Samal et al. 1994).

3.4.1
Sequence

The complete cDNA encoding the pre-B-cell CEF (PBEF_SUBDO) –related molecule, termed *SDPBEF*, was isolated (Müller et al. 1999b); it has a size of 1825 nucleotides encoding the putative 472 amino-acid-long ORF, PBEF_SUBDO (Fig. 7A). The size of the transcript is 1.9 kb and the calculated M_r of PBEF_SUBDO is 52,580.

3.4.2
Phylogenetic Relationship

Besides the human sequence no closely related molecule could be found in the databases from *D. melanogaster* or *C. elegans*. Therefore, only an unrooted tree can be presented which shows the close relationship between the sponge and human sequence (55% identical amino acids/70% similar amino acids) (Fig. 7B). In contrast, the protostomian sequences with the highest scores to the sponge sequence (*D. melanogaster*: 12% identity/29% similarity; *C. elegans*: 6% identity/22% similarity) as well as the yeast (11% identity/22% similarity) and plant sequence (14% identity/27% similarity) share only very low similarity.

3.4.3
Potential Function

In our studies to investigate if the expression of this potential chemokine also undergoes changes in the in vitro primmorph system from *S. domuncula*, these 3D-cell aggregates are exposed to self (autogeneic) as well as to nonself (xenogeneic) cell membranes from *G. cydonium* (Müller et al. 1999a). The expression studies show that the gene encoding the pre-B-cell CEF strongly increases if the primmorphs are exposed to membranes from *G. cydonium* (Fig. 4D); if membranes of *S. domuncula* were applied only a lower expression is seen. From these data, we concluded that experiments in in vitro systems are also suitable for further studies of potential chemokines.

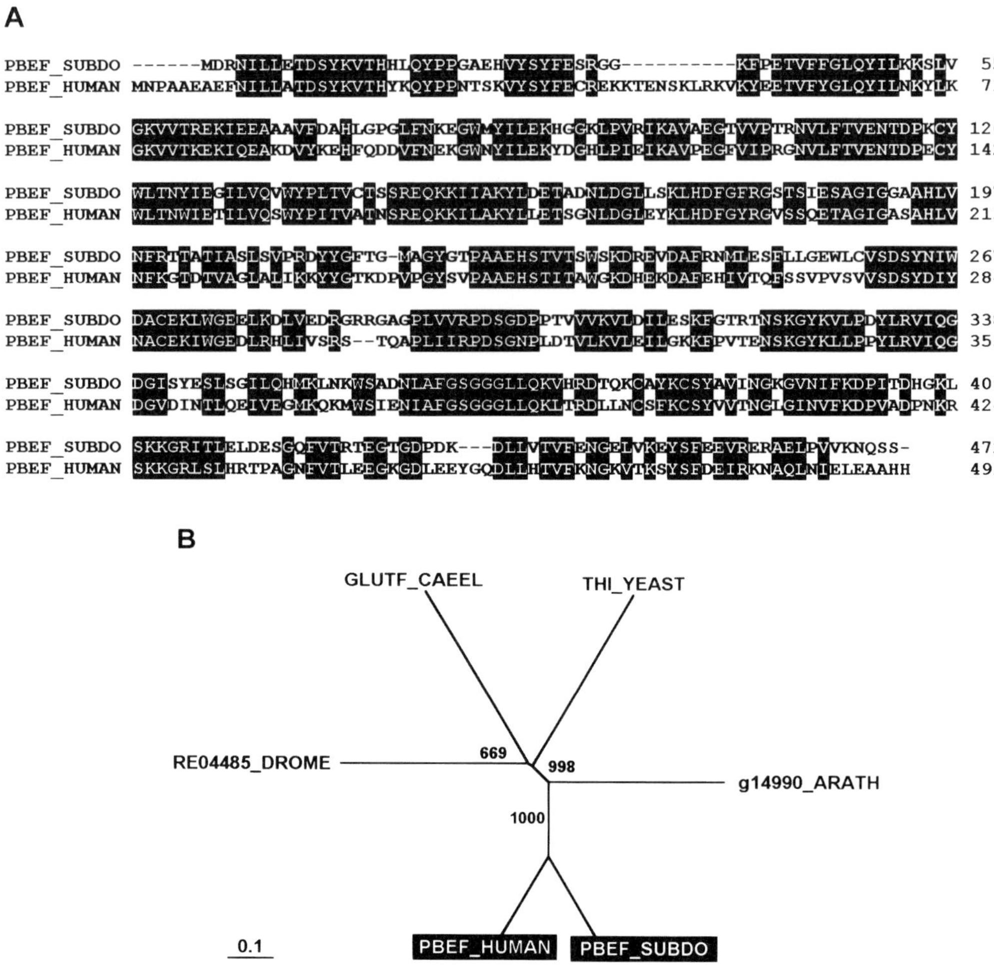

Fig. 7. The chemokine-related molecule from *S. domuncula*: the pre-B-cell colony-enhancing factor (pre-B-cell CEF). **A** The deduced aa sequence PBEF_SUBDO (Y18901) is aligned with the human pre-B-cell colony-enhancing factor (PBEF_HUMAN, NP_005737; Samal et al. 1994). **B** An unrooted tree was constructed with these two sequences and in addition with the two protostomian molecules from *D. melanogaster*, the deduced protein RE04485p (RE04485_DROME, AAN71282) and *C. elegans*, UDP-glucoronosyl and UDP-glucosyl transferase (GLUTF_CAEEL, NP_491499). Furthermore the yeast THI3 protein involved in the thiamine metabolism (THI_YEAST, NP_015065) as well as the plant g14990 unknown protein from *A. thaliana* (g14990_ARATH, AAK59544) were included

3.5 (2–5)A Synthetase

One efficient protection against invading microorganisms, primarily viruses, is the (2′–5′)oligoadenylate synthetase [(2–5)A synthetase] system (Lengyel 1982; Sen and Lengyel 1992; Rebouillat et al. 1999). The (2–5)A synthetase(s) are activated by certain classes of RNA, mainly double-stranded RNA (Hartmann et al. 1998). In addition, the (2–5)A pathway in vertebrates is induced by interferons (Pestka et al. 1987). The major enzyme of this pathway, the (2–5)A synthetase catalyzes the synthesis of a series of 2′–5′-linked oligoadenylates, termed (2–5)A [= pppA(2′p5′A)$_n$], with chain lengths of $1 \leq n \leq 30$ from ATP (Chebath et al. 1987; Hovanessian 1991). (2–5)A acts as an allosteric activator of a latent endoribonuclease, the RNase L, which degrades single-stranded, viral or cellular RNA (Zhou et al. 1993).

Recently, we have succeeded in isolating a series of cDNAs whose deduced polypeptides share high sequence similarity to (2–5)A synthetases found in vertebrates; among those were the enzymes from *G. cydonium* (Wiens et al. 1999) and from *S. domuncula* (Grebenjuk et al. 2002).

3.5.1 *Sequence*

Focusing on the synthetase from *S. domuncula*, two cDNAs, named *SD25A-1* (accession number AJ301652) and *SD25A-2* (AJ301653), have been isolated which comprise 1,175 and 1,205 nucleotides in length (Grebenjuk et al. 2002). The longest ORFs translate to 324 amino acids (for the predicted polypeptide 25A-1_SD) and to 321 amino acids (25A-2_SD), respectively (Fig. 8A). Northern blot analyses showed that the transcript length for *SD25A-1* is 1.4 kb and for *SD25A-2* 1.3 kb, indicating that the full length clones have been isolated. The calculated M_r for these new synthetases are 37,846 and 37,494, respectively.

The two *S. domuncula* sequences show the characteristic domains, found in the other (2–5)A synthetases from the sponge *G. cydonium*, mammals [mouse] and chicken: the (2–5)A synthetase signature-1 and signature-2 (Hartmann et al. 1998) (Fig. 8A). The ATP-binding site essential for enzyme activity (Ghosh et al. 1991) resides towards the C-terminus of the proteins. The dsRNA binding region of (2–5)A synthetase has been narrowed down to the segment within amino acids 105 to 163 of the mammalian enzyme (Ghosh et al. 1991). In *S. domuncula* only a very distantly related stretch exists. This finding is in agreement with a recent study which revealed that the sponge enzyme activity is not dependent on the presence of dsRNA (Lopp et al. 2002).

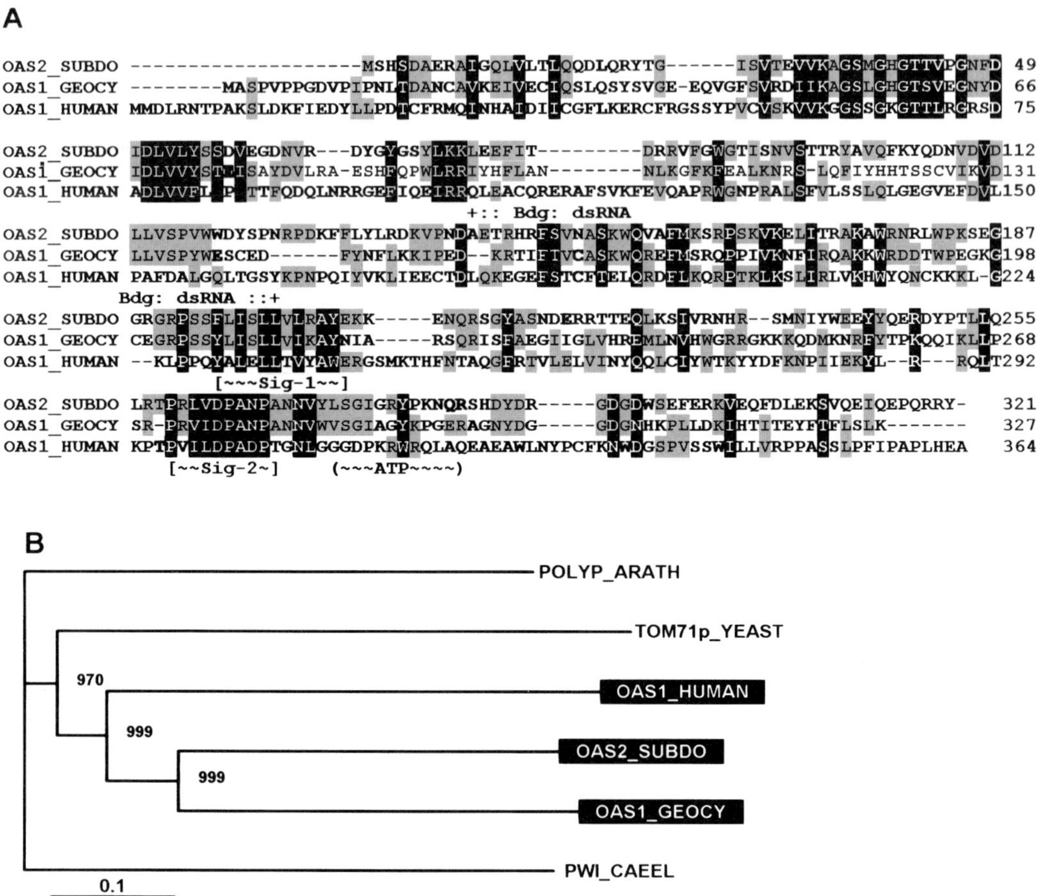

Fig. 8. Putative sponge (2–5)A synthetases from *S. domuncula* and *G. cydonium*. **A** In the alignment the *S. domuncula* enzyme [form-2] (OAS2_SUBDO; AJ301653) is compared with the *G. cydonium* enzyme (OAS1_GEOCY; Y18497) and the human OAS1 [(2–5)A synthetase, isoform E16] (OAS1_HUMAN, NP_002525; Merlin et al. 1983). Residues of aa, similar among all sequences, are in *white on black* and residues conserved in at least two sequences are in *black on grey*. The characteristic signatures of the (2–5)A synthetase are indicated: the two conserved signatures ([~Sig-1] and [~Sig-2]), the potential ATP-binding region ((~ATP)) and the dsRNA binding segment (+:: Bdg: dsRNA). **B** The rooted phylogenetic relationship of the two sponge synthetases with the very distantly related other metazoan (deuterostomian) sequence from *C. elegans*, the PWI domain containing protein (PWI_CAEEL, NP_506134), and the likewise almost unrelated molecules from yeast (the mitochondrial translocase; TOM71p_YEAST, NP_011985) and the plant *A. thaliana* (the putative retroelement; POLYP_ARATH, NP_178750) was constructed. The *A. thaliana* sequence served as outgroup. No significant protein from *D. melanogaster* was found in the database

3.5.2
Phylogenetic Relationship

The existence of the (2–5)A synthetase in several species of sponges (unpubl.) is another striking example that a series of molecules/enzymes, especially related to cytokines/chemokines (Müller et al. 2001), exists only from the Porifera to the Deuterostomia and is absent in Protostomia (Gamulin et al. 2000). Hence, the highest identities/similarities of the sponge (2–5)A synthetases exist with the vertebrate enzymes, with the human (2–5)A synthetase, isoform E16 (Merlin et al. 1983) as an example. The relationship is 18% identical amino acids and 33% similar amino acids (Fig. 8B). In contrast, the other sequences shown in this figure comprise only a low similarity (6% identity / 10% similarity).

3.5.3
Potential Function

The definite role of the (2–5)A synthetase in sponges is not known. Recently, we have performed a study with primmorphs from *S. domuncula* which were challenged with lipopolysaccharide (LPS). LPS is an endotoxin derived from the outer cell wall of Gram-negative bacteria that binds to the cell surface. By this approach, it could be demonstrated that primmorphs react to LPS treatment with a rapid increase in (2–5)A synthetase activity (Fig. 4E). In addition, it could be shown that also on the gene expression level the steady-state level of transcripts of (2–5)A synthetase gene strongly increases in primmorphs after LPS treatment (Grebenjuk et al. 2002). It is concluded that the (2–5)A synthetase system is involved in defense against microorganisms, very likely by inhibition of cell proliferation or induction of apoptosis.

4
Chemokines: Effects on Cell Proliferation

4.1
Myotrophin

The myotrophin was first found in mammalian systems. In cardiac myocytes, it stimulates protein biosynthesis (Sen et al. 1990) suggesting a crucial role in the formation of cardiac hypertrophy (reviewed in Sil et al. 1998).

4.1.1 Sequence

The cloning and expression of the *S. domuncula* myotrophin-related protein was described (Schröder et al. 2000). The cDNA, *SUBDOMYOL*, which is 582 nucleotide-long has a potential ORF of 360 nucleotides encoding a 120 amino acid-long deduced protein sequence (Fig. 9A). Northern blot analysis was performed to ensure that the sponge *SUBDOMYOL* clone is of the full length. The *S. domuncula* myotrophin, MYOL_SUBDO, has a calculated M_r of 12,837 and shares highest sequence similarity to the human myotrophin [similar to the V-1 protein; XP_059839]. The sponge sequence shows the characteristic features known from the vertebrate myotrophins (Yang et al. 1997); one half of an ankyrin repeat is located at the NH_2-terminus of the protein (amino acids 10 to 28) followed by two complete repeats towards the COOH-terminus (amino acids 30 to 62 and 63 to 95) (Fig. 9A).

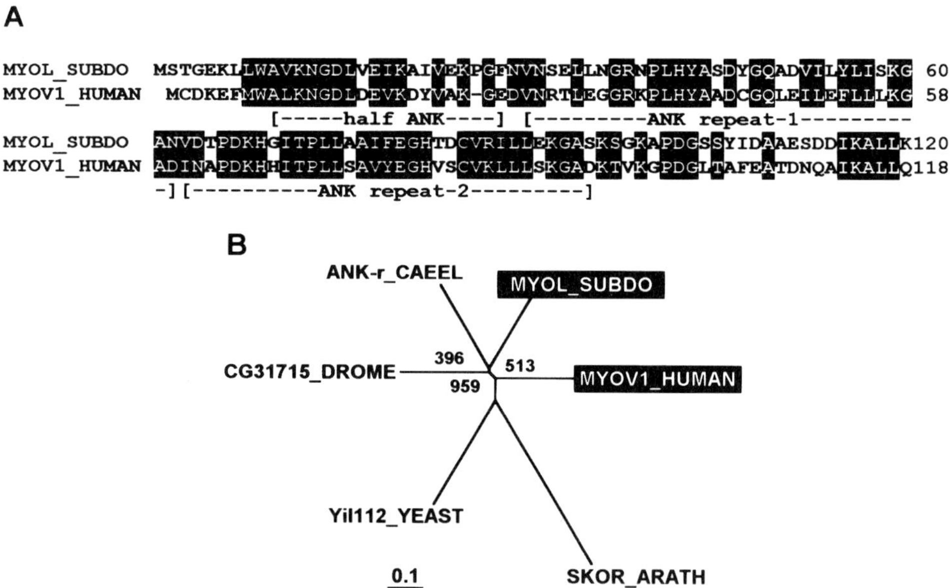

Fig. 9. Sponge myotrophin. **A** Alignment of the deduced sponge myotrophin from *S. domuncula* MYOL_SUBDO (AJ252240) with the human myotrophin, similar to the V-1 protein (MYOV1_HUMAN, XP_059839). The characteristic ankyrin motifs, one half repeat [half ANK] and two complete repeats [ANK repeat-1 and -2], are indicated. **B** Phylogenetic tree using the sponge as well as the human myotrophin and the related ankyrin sequences from *D. melanogaster* (the putative CG31715-PA protein CG31715_DROME, NP_723568) and the *C. elegans* molecule (ankyrin repeat containing protein ANK-r_CAEEL, NP_508437) as well as the yeast (hypothetical ORF from *S. cerevisiae* Yil112_YEAST, NP_012154) and plant sequence (SKOR from *A. thaliana* SKOR_ARATH, T52046). In this example the tree was not rooted to show the closer relationship between the sponge and human sequences

4.1.2
Phylogenetic Relationship

The closest similarity is seen on the basis of the sequence similarity with the human sequence (50% identical amino acids/72% similar amino acids). An unrooted phylogenetic tree was constructed to demonstrate this relationship (Fig. 9B). It shows that the next closest sequences are from *D. melanogaster*, the putative CG31715-PA protein (48% identity/67% similarity), and *C. elegans*, the ankyrin repeat containing protein (39% identity/57% similarity). Based on the amino acids within the ankyrin domains, a similarity to sequences from yeast (hypothetical ORF Yil112; 2% identity/5% similarity) and plant (SKOR from *A. thaliana*; 1% identity/3% similarity) exists.

4.1.3
Potential Function

Recombinant myotrophin was found to stimulate protein synthesis by 5-fold, as analyzed by incorporation studies using [^{3}H] lysine (Schröder et al. 2000; Fig. 10). Additionally, it was shown that after incubation of single cells with myotrophin the primmorphs show an unusual elongated, oval-shaped appearance. Furthermore, it could be demonstrated that in the presence of recombinant myotrophin the cells upregulate the expression of the collagen gene. Based

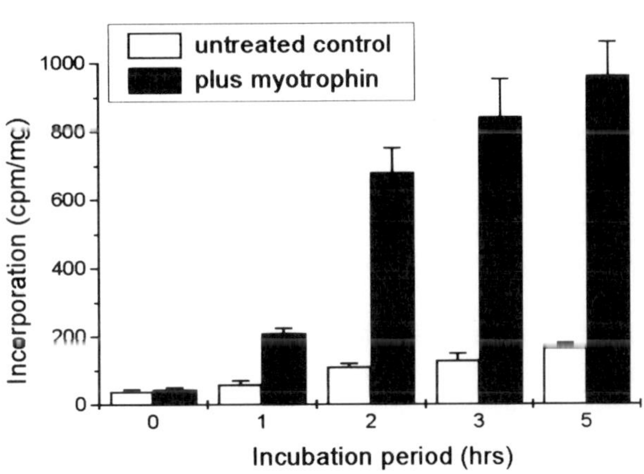

Fig. 10. Effect of recombinant myotrophin on protein synthesis in sponge cells, as measured on the basis of the incorporation of [^{3}H]Lys into acid-insoluble fractions from cells of *S. domuncula*. Single dissociated cells were incubated with [^{3}H]Lys for up to 5 h. The acid-insoluble fraction was determined and the incorporation values were correlated with the protein content in the sample. The samples remained either untreated (*open boxes*) or were incubated in the presence of 1 μg/ml of recombinant myotrophin (*closed boxes*). The results are expressed as mean values ± SD; $n = 10$

on these findings, it is concluded that the sponge myotrophin causes in homologous cells the same/similar effect as the cardiac myotrophin in mammalian cells, where it is involved in initiation of cardial ventricular hypertrophy.

5
Potential Stem Cell Marker

Stem cells are characterized by their totipotency and their self-renewal properties. Earlier experiments already showed that sponge cells contain high telomerase activity (Koziol et al. 1998; for review Müller 2003a). However, to ultimately define those cells in sponges as stem cells, marker molecules found in stem cells of triploblasts, as well as molecules allowing the differentiation of the stem cells, must be identified on gene level.

The first cDNA identified in *S. domuncula* whose deduced protein shares sequence similarity to a mammalian stem cell marker was the *S. domuncula* mesenchymal stem cell-like protein (MSCP–l).

5.1
Sponge Mesenchymal Stem Cell-Like Protein

The mesenchymal stem cell-like protein (MSCP) [accession number MN_016647] is present in mesenchymal human stem cells (van den Bos et al., unpubl.). These authors have experimental evidence that the gene encoding the MSCP is expressed in osteogenic mesenchymal stem cells. Mammalian mesenchymal stem cells are interesting since they can give rise to a range of cell types including chondrocytes, osteoblasts, adipocytes and cardiac and muscle cells (Blau et al. 2001).

5.1.1
Sequence

The *S. domuncula* mesenchymal stem cell-like protein (MSCP–l_SUBDO) was isolated by PCR. One species of cDNA was obtained with a size of 735 nucleotides (*SDMSCP-l*); Northern blot analyses confirmed that the full-length clone was obtained (Müller et al. 2003). The ORF of the clone ranges from nucleotides 48/50 to nucleotides 654/656 coding for 202 amino acids; the calculated size of MSCP-l_SUBDO is 22,535. The deduced polypeptide comprises the transmembrane helix within the amino acids 3 to 35 (Fig. 11A). One domain, present in the sequence is the consensus for the thioester superfamily (Pfam accession number PF03061); it ranges from amino acids 59 to 153 within the sponge protein.

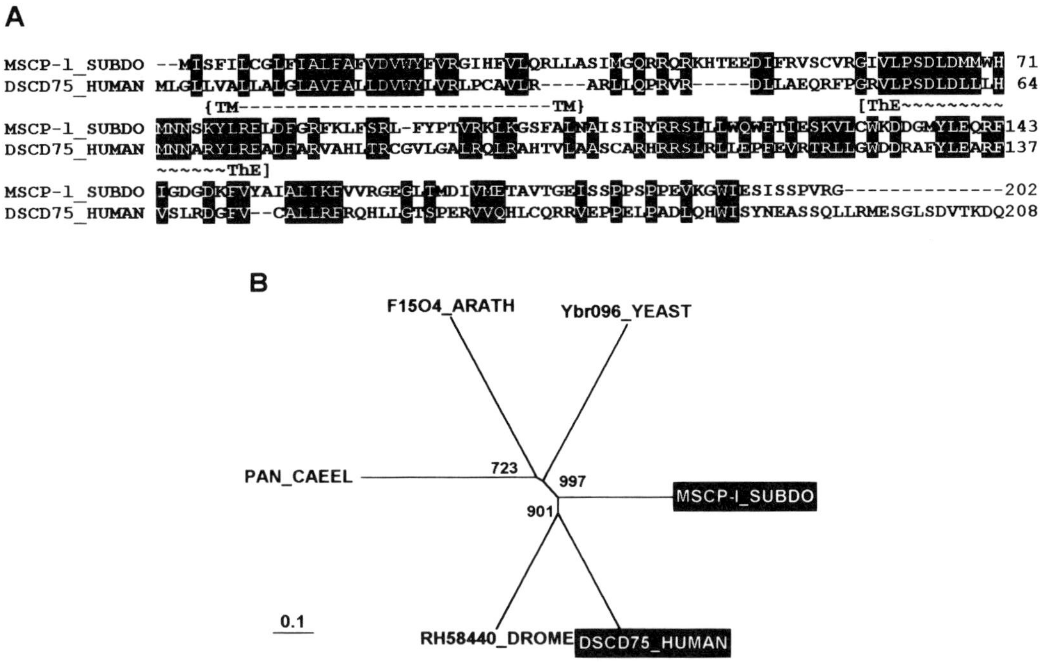

Fig. 11. *S. domuncula* mesenchymal stem cell-like protein (MSCP-l_SUBDO) as deduced from the cDNA *SUBDOMSCP-l*. **A** This protein is aligned with the human mesenchymal stem cell protein DSCD75 (DSCD75_HUMAN, AAH19104 [NIH-MGC Project URL: http://mgc.nci.nih.gov]). The consensus of the thioester superfamily ([ThE~~ThE]) as well as the putative transmembrane region ([TM–]) are marked. **B** The unrooted phylogenetic tree was constructed with these two sequences as well as the two protostomian sequences from *D. melanogaster* (the deduced RH58440p protein; RH58440_DROME, AAM52034) and *C. elegans* (PAN domain containing protein; PAN_CAEEL, NP_502253) and the *S. cerevisiae* sequence (the hypothetical ORF Ybr096wp; Ybr096_YEAST, NP_009654) and finally the plant one (the deduced protein F15O4.12; F15O4_ARATH, AAF79347)

5.1.2
Phylogenetic Relationship

Alignments were performed to estimate the phylogenetic position of the *S. domuncula* mesenchymal stem cell-like protein (MSCP-l_SUBDO). Again this sequence shares the highest similarity to the human sequence (mesenchymal stem cell protein DSCD75) with 27% identical amino acids and 46% related amino acids. The scores for the *D. melanogaster* enzyme (23% identity/41% similarity) are lower. Very little similarity is seen after comparison with the sequences of *C. elegans* (5% identity/9% similarity), *S. cerevisiae* (17% identity/33% similarity) and *A. thaliana* (5% identity/11% similarity; Fig. 11B).

5.1.3 Potential Function

Until now, only one first piece of functional evidence about the role of the MSCP-l is available (Müller et al. 2003). It has been described that the expression of *MSCP-l* is under control of the morphogenetic inorganic elements, silicon and ferric iron. Northern blot studies revealed that the potency of the sponge cells in primmorphs to proliferate is paralleled by an increased expression of the *S. domuncula MSCP-l* gene (Fig. 12). However, if the primmorphs were cultivated in the presence of 60 μM silicate and 30 μM Fe^{3+}, the amount of transcripts of this gene had a lower level 1.3-fold [minus silicate/Fe^{3+}] versus 0.4-fold [plus silicate/Fe^{3+}]. The notion that the MSCP-l from *S. domuncula* serves also in sponges as a stem cell marker might be supported by the expression studies of this gene. Based on earlier studies, it can be assumed that during spicule formation in primmorphs which is dependent on silicate/Fe^{3+}, the amount of spicule-forming and collagen-synthesizing cells increases (Krasko et al. 2000) on expense of the totipotent archaeocytes.

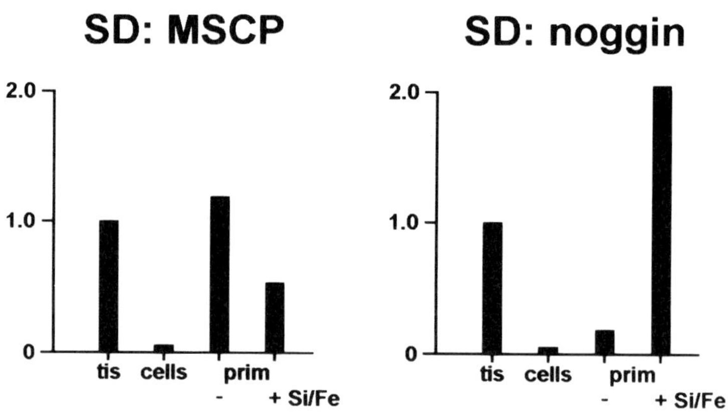

Fig. 12. Level of transcripts for genes potentially indicative for stem cells in tissue or primmorphs from *S. domuncula*. Tissue samples (*tis*) were obtained from animals which were kept in aquaria. Primmorphs (*prim*) formed after five days remained either untreated and were kept in natural seawater, in the absence (–Si/Fe) or presence of 60 μM silicate and 30 μM Fe^{3+} (+Si/Fe). In one experiment, single cells were used which were kept in Ca^{2+} and Mg^{2+}-free artificial seawater for 3 days (cells). Northern blot analyses were performed with the probe for the mesenchymal stem cell-like protein, *SDMSCP-l*, or the noggin probe *SDNOGGIN*. The intensities of the bands were quantified and given as relative degree of expression; the expression in sponge tissue was set to onefold

6 Chemokines Potentially Involved in Differentiation of Stem Cells

Two genes have recently been identified which encode proteins potentially involved in stem cell maturation, noggin and the glia maturation factor.

6.1 Noggin

Noggin was initially isolated from *Xenopus* (Smith and Harland 1992); it is expressed in the (Spemann's) organizer (Lemaire and Kodjabachian 1996). In vertebrates, noggin is involved in the formation of dorsal mesoderm derivatives, e.g. the skeletal muscles (Smith et al. 1993). The first noggin protein/cDNA from invertebrates has been identified in a planarian, where it is very likely involved in the blastema formation as well as in dorsoventral patterning (Ogawa et al. 2002).

6.1.1 *Sequence*

The noggin-like protein from *S. domuncula* (NOGG-l_SUBDO) was deduced from the 941 nucleotide-long *SDNOGG-l* cDNA. The ORF spans from nucleotides 5/7 to nucleotides 704/706. The 234 amino acids of the sponge noggin have a size of 26,459 kDa and comprise the characteristic Cys residues which are marked in Fig. 13A.

6.1.2 *Phylogenetic Relationship*

The human noggin protein (Valenzuela et al. 1995) is closely related to the sponge noggin protein NOGG-l_SUBDO with 25% identical amino acids and 36% similar amino acids. The protostomian sequence from *D. melanogaster* (the Trunk protein) is only related with a score of 16% identical amino acids and 24% similar amino acids, and from *C. elegans* (SPE-6 protein kinase), only with 15% identical amino acids and 25% similar amino acids. Even lower is the similarity to *S. cerevisiae* (the ATP-dependent helicase), 6% identical amino acids and 16% related amino acids, and to *A. thaliana* (a protein kinase-like protein) 6% identical amino acids and 12% similar amino acids (Fig. 13B).

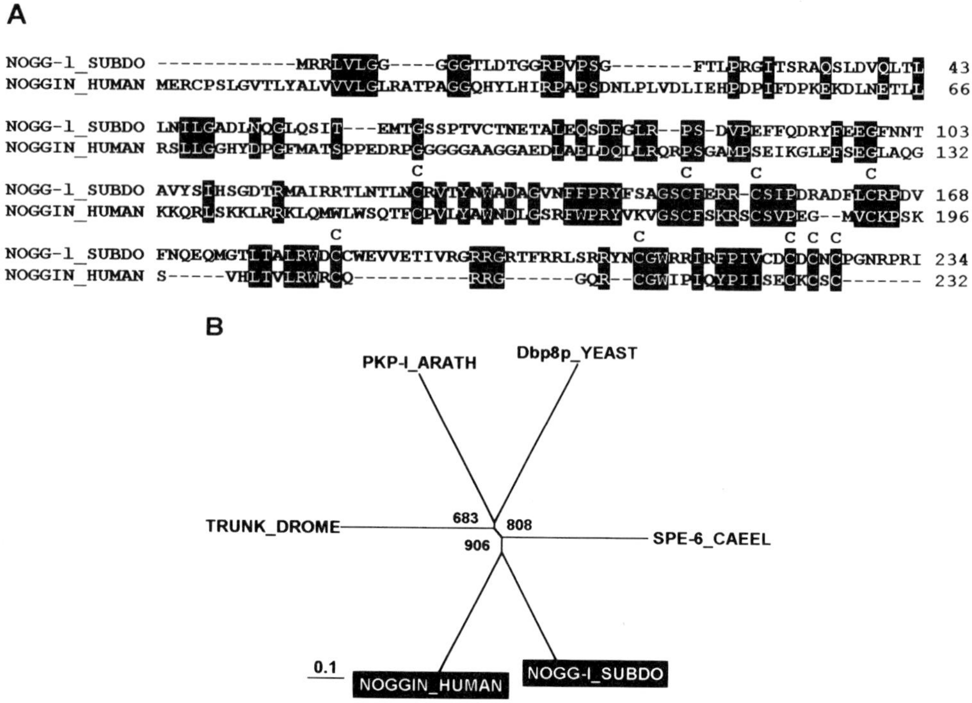

Fig. 13. Noggin protein from *S. domuncula* (NOGG-l_SUBDO) is aligned (**A**) with the related human noggin precursor (NOGGIN_HUMAN, NP_005441; Valenzuela et al. 1995). The conserved Cys residues are marked above the sequences. **B** The unrooted tree is constructed with these sequences and the related ones from *D. melanogaster* (Trunk protein; TRUNK_DROME, Q24155), *C. elegans* (SPE-6 protein kinase; spe-6_CAEEL, NP_499482), *S. cerevisiae* (the ATP-dependent helicase; Dbp8p_YEAST, NP_012039) and *A. thaliana* (protein kinase-like protein; PKP-l_ARATH, NP_190952)

6.1.3
Potential Function

Expression studies of the noggin gene were performed. They revealed that the transcript level of SDNOGG-l is high in tissue in comparison to the expression in primmorphs which were kept in the absence of silicate/Fe^{3+} (tissue versus primmorphs 1:0.2-fold). When primmorphs were cultivated in the presence of 60 μM silicate and 30 μM Fe^{3+}, expression of noggin increased to 2.3-fold in primmorphs (Fig. 12). This finding was taken as an indication that during morphogenetic events, e.g. spicule formation, noggin is upregulated. In view of the data observed in vertebrates (Gratsch and O'Shea 2002), it is assumed that also in sponges this factor acts as an inducer of differentiation.

6.2 Glia Maturation Factor

The glia maturation factor is a small protein which is restricted in vertebrates to the nervous system (Lim et al. 1989; Zaheer et al. 1993). Experimental evidence is available indicating that this protein exerts both an extracellular effect (Nieto-Sampedro et al. 1988) and an intracellular role (Lim et al. 2000). Extracellularly, the factor has been assumed to be involved in healing responses (Nieto-Sampedro et al. 1988), while intracellularly, it acts as a signaling molecule to promote survival and resistance to apoptosis (Lim et al. 2000).

6.2.1 Sequence

The complete sponge sequence comprises a 759 nucleotide-long cDNA. The ORF spans from nucleotides 131 to 571, and encodes a 147 amino acid-long deduced protein. One conserved domain exists, which is termed ADF/cofilin consensus (Hawkins et al. 1993), is indicated in Fig. 14A.

6.2.2 Phylogenetic Relationship

The *S. domuncula* glia maturation factor-like protein shows the following sequence relationships: to the human factor 52% identical amino acids and 67% similar amino acids, the *D. melanogaster* molecule 46% identical amino acids and 65% similar amino acids, the *C. elegans* protein 35% identical amino acids and 65% similar amino acids, the *S. cerevisiae* hypothetical ORF 26% identical amino acids and 43% similar amino acids, and the plant factor 20% identical amino acids and 39% similar amino acids (Fig. 14B).

6.2.3 Potential Function

The function of the glia maturation factor-like protein in *S. domuncula* is under investigation.

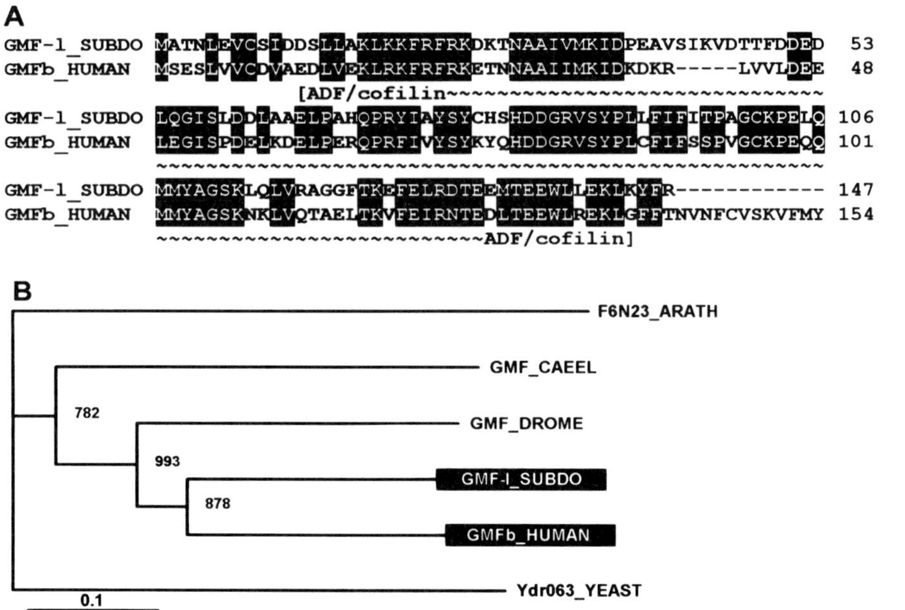

Fig. 14. Glia maturation factor-like protein from *S. domuncula* (GMF-l_SUBDO) is aligned with the human glia maturation factor beta (GMFb_HUMAN, AAH05359) (**A**). The conserved domain ADF/cofilin is marked. **B** The phylogenetic tree is constructed with the related molecules from *D. melanogaster* (GMF_DROME, NP_609787), *C. elegans* (GMF_CAEEL, AAN84814), *S. cerevisiae* (hypothetical ORF Ydr063wp; Ydr063_YEAST, NP_010348) and the *A. thaliana* actin-depolymerizing factor F6N23.12 (F6N23_ARATH, T01232). The tree is rooted with the plant sequence

7
Sponges as Complex and Integrated Animals: Immune System and Stem Cell System as Key Novelties with Chemokines as Mediators

It is now conclusively established that the common ancestor of the Metazoa, the hypothetical Urmetazoa, shares also a common ancestor with the fungi (Schütze et al. 1999). Nevertheless, the Metazoa have developed genuine novelties, which are restricted to this kingdom only. Among those molecules are the adhesion molecules and the components associated with them (Müller 2001; Müller et al. 2001).

In an earlier review (Müller et al. 1999b), which is now extended here, it could be demonstrated that the immune system within the Metazoa has likewise a common pattern; especially the chemokines share common features within Metazoa. It is outlined here that the so far known (putative) chemokines identified in sponges share high sequence similarity to the related molecules present in triploblasts. Only on a first glance, it seemed surprising that most of the sponge molecules identified are on protein sequence level closer related to Deuterostomia than to Protostomia. Good examples for this fact are the chemokines, or enzymes producing mediators, summarized here: the AIF, the endothelial-monocyte-activating polypeptide, the pre-B-cell CEF or the (2–5)A synthetases. Considering the fact that the diploblastic organisms precede the emergence of the Protostomia and Deuterostomia, it can now be stated that they comprised already the basic elements of the metazoan immune system; some of the pathways in common between Porifera and Deuterostomia have been lost in Protostomia; an example is the (2–5)A synthetases system. For some of the factors in sponges, a close functional relationship as immune molecules could be demonstrated, e.g. for the myotrophin, AIF or the (2–5)A synthetase; the genes encoding these molecules are upregulated, or their activities increase after contact to nonself tissue/cells.

As outlined recently (Müller 2003b), the first steps required for the establishment of the colonial system was the acquisition of the molecules involved in cell – cell adhesion which allows the transfer of signals between cell assemblies. The consequences of this level of integration are activation of intracellular signal transduction pathways that result in differential gene expression and cell specialization. Secondly, cell – matrix adhesion molecules had to be invented that allow the integration of the functional units.

These adhesion systems are the basis for reaching the colonial stage of the metazoans and prerequisites for the establishment of integrated systems (Fig. 15). These adhesion systems alone are not sufficient for individuation. The stage of individuation can only be reached after the acquisition of an immune system which is paralleled or complemented by a mechanism that eliminates unwanted cells which accumulate during development. In a multicellular organism, this process is termed apoptosis. Here we highlight again, that sponges have an amazingly complex immune system, which acts against invading microbes or parasites. In parallel with the evolution of the molecules characteristic for the metazoan immune system, the enzymatic as well as structural elements for the process of apoptosis were also formed; the metazoan prototype molecules involved in apoptosis are again found in Porifera (e.g. Wiens et al. 2000a,b, 2001). Apoptosis is the mechanism in Metazoa that guarantees homeostasis.

One evident difference between diploblasts and triploblasts is the absence of any vascular system required for the "transport" of the cytokines/chemokines in a soluble form. Hence, these diploblastic animals had to develop a different strategy to position these factors to their functional destination. Focusing on sponges, the chemokines had to be transported through the mesohyl, a very flexible and dynamic matrix into which the cells are embedded. Evidence has been presented that calcium ions are one factor which controls the gel–sol property of the extracellular matrix molecules, with the galectins as the prominent molecule (Müller et al. 1983). The state of the cells in the mesohyl of the sponges, not connected to organs, allows an efficient

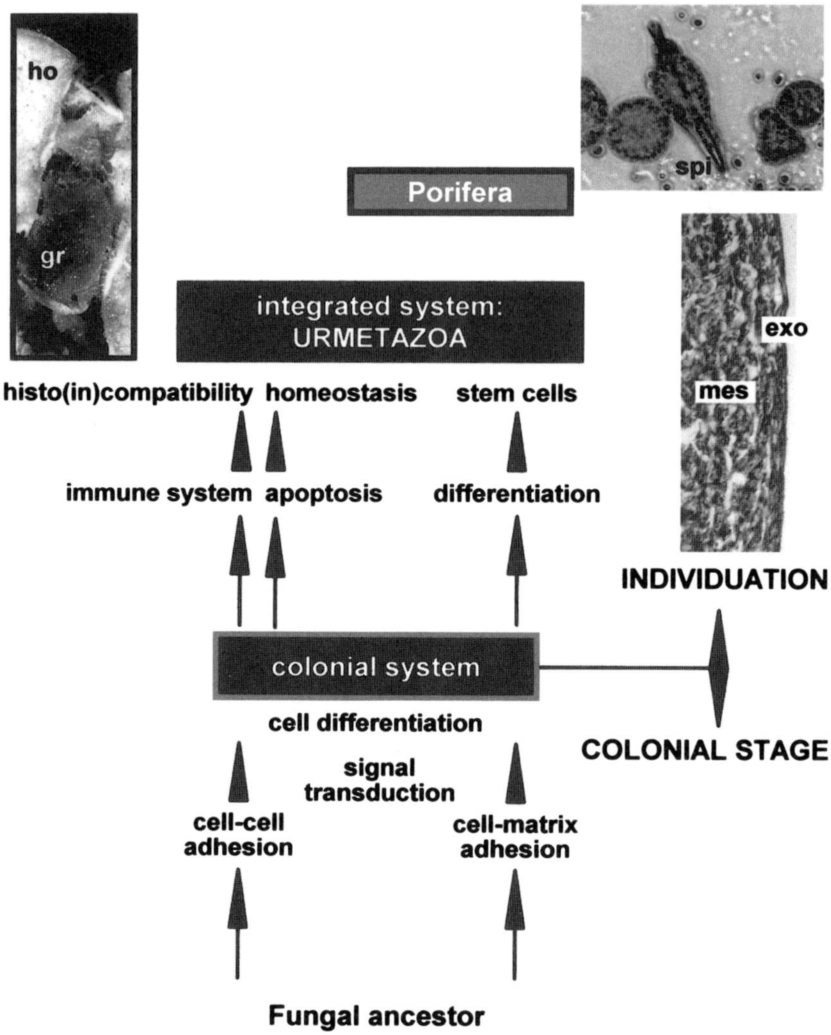

Fig. 15. Hypothetical steps towards the evolution to the Urmetazoa with the Porifera as the next closest taxon. Adhesion molecules were required to allow the transition from a fungal-like ancestor to a colonial system. This stage, which made the further development of immune- and apoptotic systems possible, led to the evolution of the Urmetazoa as an integrated system. The sponges have developed already efficient strategies that allowed the rejection of allogeneic tissue/grafts (*gr*) by the host (*ho*), here with *G. cydonium* as a model. As a further prerequisite of the transition of the colonial ancestor to the Urmetazoa, as an integrated animal system, the compartmentalization of the cells into differentiated somatic cells, like the exopinacocytes (*exo*) present as the surface layer around the mesohyl (*mes*) or the spicules-forming sclerocytes (*spi*), and stem cells had to be established (*S. domuncula*)

transport of the chemokines to their target cells. Only the pinacoderm/choanoderm areas in the sponges, with the choanocyte chambers, can be considered as organ-like cell assemblies since there the cells are firmly attached and remain in a more static position.

However, a functional immune/apoptotic system is also not sufficient to allow the stage of individuation. This level can be reached only by the formation of differentiated cells under maintenance of some totipotent stem cells. Hence, with the establishment of a Bauplan in Porifera through the complex network of transcription factors, some cells must have retained their self-renewal totipotent capacity as the population of stem cells. It has been proven that most of the sponge cells, if they are present in a close functional connection with other cells, are rich in telomerase activity (for review Müller 2003a). Here, in this review the first genes/molecules are described which are characteristic for stem cells of triploblasts, e.g. the mesenchymal stem cell-like protein, or the "organizer" molecules, e.g. noggin. It might be allowed to point out here that it was indeed a surprise that a series of molecules/genes could be identified in sponges which are known to exist in the Spemann's organizer of vertebrates (Spemann 1936; Lemaire and Kodjabachian 1996).

The pinacocytes (epithelial cells), the myocytes (contractile cells) and the sclerocytes/collencytes (skeletal cells) represent in sponges the differentiated cells originating form the totipotent archaeocytes (Fig. 1). The self-renewal potency is retained in stem cells which allow the repair and replacement of these somatic cells (Fig. 15). Based on the data given in this review, solid ground is now available for the assumption of a common immune system, the histo(in)compatibility system which allows the homeostasis within an individual, in all metazoan phyla. In parallel with the establishment of a complex organism in sponges built from differentiated cells, a defined population of cells also exists which retain the totipotent self-renewal capacity, the stem cells. With the demonstration of stem cell markers, as well as factors allowing their differentiation, a new avenue of research has been opened to understand the evolutionary transitions to the Urmetazoa. Through the analysis of the genome of sponges in general and of the sponges *S. domuncula* and *G. cydonium* in particular sponges are a fascinating experimental model which allows to stethoscope into the evolutionary mechanisms which proceeded approximately 600 to 800 MYA.

Acknowledgement. This work was supported by the Deutsche Forschungsgemeinschaft, the International Human Frontier Science Program [RG-333/96-M] and the Bundesministerium für Bildung und Forschung (project: Center of Excellence *Biotec*marin).

References

Adell T, Wiens M, Müller WEG (2002) SD-Bra and SD-Tbx: first T-box transcription factors from Porifera. Boll Mus Ist Biol Univ Genova 66/67:5–6

Blumbach B, Pancer Z, Diehl-Seifert B, Steffen R, Münkner J, Müller I, Müller WEG (1998) The putative sponge aggregation receptor: isolation and characterization of a molecule composed of scavenger receptor cysteine-rich domains and short consensus repeats. J Cell Sci 111:2635-2644

Blau HM, Brazelton TR, Weimann JM (2001) The evolving concept of a stem cell: entity or function? Cell 105:829–841

Buccich G (1886) Alcune spugne dell'Adriatico sconosciute e nuove. Boll Soc Adriat Sci Nat Trieste 9:222–225

Burger MM, Jumblatt J (1977) Membrane involvement in cell-cell interactions: a two component model system for cellular recognition that does not require live cells. In: Lash JW, Burger MM (eds) Cell and tissue interactions. Raven Press, New York, pp 155–172

Buscema M, van de Vyver G (1983) Variability of allograft rejection processes in *Axinella verrucosa*. Dev Comp Immunol 7:613–616

Cauldwell C, Henkart P, Humphreys T (1973) Physical properties of sponge aggregation factor: a unique proteoglycan complex. Biochemistry 12:3045–3055

Chebath J, Benech P, Hovanessian AG, Galabru J, Robert N, Revel M (1987) Four different forms of interferon-induced 2′,5′-oligo(A) synthetase identified by immunoblotting in human cells. J Biol Chem 262:3852–3857

Chen ZW, Ahren B, Östenson CG, Cintra A, Bergman T, Müller C, Fuxe K, Mutt V, Jörnvall H, Efendic S (1997) Identification, isolation, and characterization of daintain (allograft inflammatory factor 1), a macrophage polypeptide with effects on insulin secretion and abundantly present in the pancreas of prediabetic BB rats. Proc Natl Acad Sci USA 94:13879–13884

Cikryt P, Feuerstein S, Wendel A (1982) Selenium- and non-selenium-dependent glutathione peroxidases in mouse liver. Biochem Pharmacol 31:2873–2877

Custodio MR, Prokic I, Steffen R, Koziol C, Borojevic R, Brümmer F, Nickel M, Müller WEG (1998) Primmorphs generated from dissociated cells of the sponge *Suberites domuncula*: a model system for studies of cell proliferation and cell death. Mech Ageing Dev 105:45–59

Diehl-Seifert B, Kurelec B, Zahn RK, Dorn A, Jerecevic B, Uhlenbruck G, Müller WEG (1985) Attachment of sponge cells to collagen substrata: effect of a collagen assembly factor. J Cell Sci 79:271–285

Fernandez-Busquets X, Burger MM (1997) The main protein of the aggregation factor responsible for species-specific cell adhesion in the marine sponge *Microciona prolifera* is highly polymorphic. J Biol Chem 272:27839–27847

Fernandez-Busquets X, Burger MM (1999) Cell adhesion and histocompatibility in sponges. Microsc Res Techn 44:204–218

Fernandez-Busquets X, Kammerer RA, Burger MM (1996) A 25 kDa protein is the basic unit of the core from the 2×10^4-kDa aggregation factor responsible for species-specific cell adhesion in the marine sponge *Microciona prolifera*. J Biol Chem 271:23558–23565

Galtsoff PS (1925) Regeneration after dissociation (an experimental study on sponges) II. J Exp Zool 42:223–251

Gamulin V, Müller IM, Müller WEG (2000) Sponge proteins are more similar to those of *Homo sapiens* than to *Caenorhabditis elegans*. Biol J Linnean Soc 71:821–828

Garrone R (1978) Phylogenesis of connective tissue. Morphological aspects and biosynthesis of sponge intercellular matrix. Karger, Basel

Garrone R (1998) Evolution of metazoan collagens. Prog Mol Subcell Biol 21:119–139

Ghosh SK, Kusari J, Bandyopadhyay SK, Samanata H, Kumar R, Sen GC (1991) Cloning, sequencing, and expression of two murine 2′-5′-oligoadenylate synthetases. J Biol Chem 266:15293–15299

Gonzales JM, Moran MA (1997) Numerical dominance of a group of marine bacteria in the alpha-subclass of the class *Proteobacteria* in coastal seawater. Appl Environ Microbiol 63: 4237–4242

Gratsch TE, O'Shea KS (2002) Noggin and chordin have distinct activities in promoting lineage commitment of mouse embryonic stem (ES) cells. Dev Biol 245:833–894

Grebenjuk VA, Kuusksalu A, Kelve M, Schütze J, Schröder HC, Müller WEG (2002) Induction of (2′–5′)oligoadenylate synthetase in the marine sponges *Suberites domuncula* and *Geodia cydonium* by the bacterial endotoxin lipopolysaccharide. Eur J Biochem 269:1382–1392

Hartmann R, Noerby PL, Martensen PM, Joergensen P, James MC, Jacobson C, Moestrup SK, Clemens MJ, Justesen J (1998) Activation of 2-5 oligoadenylate synthetase by single-stranded and double-stranded RNA aptamers. J Biol Chem 273:3236–3246

Hawkins M, Pope B, Maciver SK, Weeds AG (1993) Human actin depolymerizing factor mediates a pH-sensitive destruction of actin filaments. Biochemistry 32:9985–9993

Henkart P, Humphreys S, Humphreys T (1973) Characterization of sponge aggregation factor. A unique proteoglycan complex. Biochemistry 12:3045–3050

Hildemann WH, Johnston IS, Jokiel PL (1979) Immunocompetence in the lowest metazoan phylum: transplantation immunity in sponges. Science 204:420–422

Hirabayashi J, Kasai K (1993) The family of metazoan metal-independent ß-galactoside-binding lectins: structure, function and molecular evolution. Glycobiology 3:297–304

Hoshiyama D, Suga H, Iwabe W, Koyanagi M, Nikoh N, Kuma K, Matsuda F, Honjo T, Miyata T (1998) Sponge Pax cDNA related to Pax-2/5/8 and ancient gene duplications in the Pax family. J Mol Evo 47:640–648

Hovanessian AG (1991) Interferon-induced and double-stranded RNA-activated enzymes: a specific protein kinase and 2′,5′-oligoadenylate synthetases. J Interferon Res 11:199–205

Humphreys T (1963) Chemical dissociation and in vitro reconstruction of sponge cell adhesion I. Dev Biol 8:27–47

Humphreys T (1994) Rapid allorecognition in the marine sponge *Microciona prolifera*. Implications for evolution of immune recognition. Ann NY Acad Sci 712:342–345

Imai Y, Ibata I, Ito D, Ohsawa K, Kohsaka S (1996) A novel gene *iba1* in the major histocompatibility complex class III region encoding an EF hand protein expressed in a monocytic lineage. Biochem Biophys Res Commun 224:855–862

Iris F, Bougueleret L, Prieur S, Caterina D, Primas G, Perrot V, Jurka J, Rodriguez-Tome P, Claverie J, Cohen D, Dausset J (1993) Dense Alu clustering and a potential new member of the NF kappa B family within a 90 kilobase HLA class III segment. Nature Genet 3:137–145

Kao J, Ryan J, Brett J, Chen J, Shen H, Fan Y, Godman G, Familletti PC, Wang F, Pan YCE, Clauss M (1992) Endothelial monocyte-activating polypeptide II; a novel tumor-derived polypeptide that activates host-response mechanisms. J Biol Chem 267:20239–20247

Kao J, Houck K, Fan Y, Haehnel I, Libutti SK, Kayton ML, Grikscheit T, Chabot J, Nowygrod R, Greenberg S, Kuang WJ, Leung DW, Hayward JR, Kisiel W, Heath M, Brett J, Stern DM (1994a) Characterization of a novel tumor-derived cytokine. Endothelial-monocyte-activating polypeptide II. J Biol Chem 269:25109–25119

Kao J, Fan Y, Haehnel I, Brett J, Greenberg S, Clauss M, Kayton ML, Houck K, Kisiel W, Seljelid R, Burnier J, Stern DM (1994b) A peptide derived from the amino terminus of endothelial-monocyte-activating polypeptide II modulates mononuclear and polymorphonuclear leukocyte functions, defines an apparently novel cellular interaction site, and induces an acute inflammatory response. J Biol Chem 269:9774–9782

Kheradmand F, Werner E, Tremble P, Symons M, Werb Z (1998) Role of *Rac1* and oxygen radicals in collagenase-1 expression induced by cell shape change. Science 280:898–902

Kim TS, Sundaresh CS, Feinstein SI, Dodia C, Skach WR, Jain MK, Nagase T, Seki N, Ishikawa K, Nomura N, Fisher AB (1997) Identification of a human cDNA clone for lysosomal type Ca^{2+}-independent phospholipase A2 and properties of the expressed protein. J Biol Chem 272: 2542–2550

Koziol C, Borojevic R, Steffen R, Müller WEG (1998) Sponges (Porifera) model systems to study the shift from immortal- to senescent somatic cells: the telomerase activity in somatic cells. Mech Ageing Develop 100:107–120

Krasko A, Batel R, Schröder HC, Müller IM, Müller WEG (2000) Expression of silicatein and collagen genes in the marine sponge *Suberites domuncula* is controlled by silicate and myotrophin. Eur J Biochem 267:4878–4887

Krasko A, Schröder HC, Batel R, Grebenjuk VA, Steffen R, Müller IM, Müller WEG (2002) Iron induces proliferation and morphogenesis in primmorphs from the marine sponge *Suberites domuncula*. DNA Cell Biol 21:67–80

Kruse M, Steffen R, Batel, R, Müller IM, Müller WEG (1999) Differential expression of allograft inflammatory factor 1 and of glutathione peroxidase during auto- and allograft response in marine sponges. J Cell Sci 112:4305–4313

Laurent PL (1842) Recherches sur l'hydre et l'éponge d'eau douce. Bertrand, Paris

Le Pennec G, Perovic S, Ammar MSA, Grebenjuk VA, Steffen R, Brümmer F, Müller WEG (2003) Cultivation of primmorphs from the marine sponge *Suberites domuncula*: morphogenetic potential of silicon and iron. A review. Marine Biotechnol (in press)

Lemaire P, Kodjabachian L (1996) The vertebrates organizer: structure and molecules. Trends Genet 12:525–531

Lengyel PARB (1982) Biochemistry of interferons and their action. Annu Rev Biochem 51:251–282

Lim R, Miller JF, Zaheer A (1989) Purification and characterization of glia maturation factor beta: a growth regulator for neurons and glia. Proc Natl Acad Sci USA 86:3901–3905

Lim R, Zaheer A, Yorek MA, Darby CJ, Oberley LW (2000) Activation of nuclear factor-kappa B in C6 rat glioma cells after transfection with glia maturation factor. J Neurochem 74:596–602

Lopp A, Kuusksalu A, Reintamm T, Müller WEG, Kelve M (2002) 2'5'-oligoadenylate synthetase from a lower invertebrate, the marine sponge *Geodia cydonium*, does not need dsRNA for its enzymatic activity. Biochim Biophys Acta 1590:140–149

Manuel M, Le Parco Y (2000) Homeobox gene diversification in the calcareous sponge, *Sycon raphanus*. Mol Phylog Evol 17:97–107

Merlin G, Chebath J, Benech P, Metz R, Revel M (1983) Molecular cloning and sequence of partial cDNA for interferon-induced (2′-5′)oligo(A) synthetase mRNA from human cells. Proc Natl Acad Sci USA 80:4904–4908

Misevic GN, Jumblatt JE, Burger MM (1982) Cell binding fragments from a sponge proteoglycan-like aggregation factor. J Biol Chem 257:46931–46936

Moore HF (1908) The commercial sponges and the sponge fisheries. Bull Bureau Fisheries 28: 399–511

Mukai H (1992) Allogeneic recognition and sex differentiation in chimeras of the freshwater sponge *Ephydatia muelleri*. J Exp Zool 264:298–311

Müller WEG (1982) Cell membranes in sponges. Int Rev Cytol 77:129–181

Müller WEG (1995) Molecular phylogeny of Metazoa (animals): monophyletic origin. Naturwissenschaften 82:321–329

Müller WEG (1997) Origin of metazoan adhesion molecules and adhesion receptors as deduced from their cDNA analyses from the marine sponge *Geodia cydonium*. Cell Tissue Res 289:3 83–395

Müller WEG (2001) How was metazoan threshold crossed: the hypothetical Urmetazoa. Comp Biochem Physiol [A] 129:433–460

Müller WEG (2003a) Telomerase activity in sponges (Porifera) the closest related taxa of the hypothetical ancestral animal the Urmetazoa. In: Krupp G, Parwaresch R (eds) Telomerases, telomeres and cancer, Molecular Biology Intelligence, Unit 22. Landes Bioscience/Eureka, Georgetown, TX, pp 300–313

Müller WEG (2003b) The origin of metazoan complexity: Porifera as integrated animals. Integ Comp Biol (in press)

Müller WEG, Schäcke H (1996) Characterization of the receptor protein-tyrosine kinase gene from the marine sponge *Geodia cydonium*. Prog Mol Subcell Biol 17:183–208

Müller WEG, Zahn RK (1973) Purification and characterization of a species-specific aggregation factor in sponges. Exp Cell Res 80:95–104

Müller WEG, Müller I, Zahn RK, Kurelec B (1976) Species-specific aggregation factor in sponges. VI. Aggregation receptor from the cell surface. J Cell Sci 21:227–241

Müller WEG, Zahn RK, Arendes J, Kurelec B, Steffen R, Müller I (1979) Aggregation of sponge cells. Self-aggregation of the circular proteid particle. Biochim Biophys Acta 551:363–367

Müller WEG, Conrad J, Schröder C, Zahn RK, Kurelec B, Dreesbach K, Uhlenbruck G (1983) Characterization of the trimeric, self-recognizing *Geodia cydonium* lectin I. Eur J Biochem 133: 263–267

Müller WEG, Wiens M, Batel R, Steffen R, Borojevic R, Custodio MR (1999a) Establishment of a primary cell culture from a sponge: primmorphs from *Suberites domuncula*. Marine Ecol Prog Ser 178:205–219

Müller WEG, Perovic S, Wilkesman J, Kruse M, Müller IM, Batel R (1999b) Increased gene expression of a cytokine-related molecule and profilin after activation of *Suberites domuncula* cells with xenogeneic sponge molecule(s). DNA Cell Biol 18:885–893

Müller WEG, Schröder HC, Skorokhod A, Bünz C, Müller IM, Grebenjuk VA (2001) Contribution of sponge genes to unravel the genome of the hypothetical ancestor of Metazoa (Urmetazoa). Gene 276:161–173

Müller WEG, Krasko A, Skorokhod A, Bünz C, Grebenjuk VA, Steffen R, Batel R, Müller IM, Schröder HC (2002) Histocompatibility reaction in the sponge *Suberites domuncula* on tissue and cellular level: central role of the allograft inflammatory factor 1. Immunogenetics 54: 48–58

Müller WEG, Korzhev M, Le Pennec G, Müller IM, Schröder HC (2003) Origin of metazoan stem cell system in sponges: first approach to establish the model (*Suberites domuncula*). Biomol Eng (in press)

Munz B, Frank S, Huebner G, Olsen E, Werner S (1997) A novel type of glutathione peroxidase: expression and regulation during wound repair. Biochem J 326:579–585

Nieto-Sampedro M, Lim R, Hicklin DJ, Cotman CW (1988) Early release of glia maturation factor and acidic fibroblast growth factor after rat brain injury. Neurosci Lett 86:361–365

Ogawa K, Ishihara S, Saito Y, Mineta K, Nakazawa M, Ikeo K, Gojobori T, Watanabe K, Agata K (2002) Induction of a *noggin*-like gene by ectopic DV interaction during planarian regeneration. Dev Biol 250:59–70

Ognjanovic S, Bao S, Yamamoto SY, Garibay-Tupas J, Samal B, Bryant-Greenwood GD (2001) Genomic organization of the gene coding for human pre-B-cell colony-enhancing factor and expression in human fetal membranes. J Mol Endocrinol 26:107–117

Pahler S, Krasko A, Schütze J, Müller IM, Müller WEG (1998) Isolation and characterization of a cDNA encoding a potential morphogen from the marine sponge *Geodia cydonium*, that is conserved in higher metazoans. Proc R Soc Lond B 265:421–425

Pancer Z, Kruse M, Schäcke H, Scheffer U, Steffen R, Kovács P, Müller WEG (1996) Polymorphism in the immunoglobulin-like domains of the receptor tyrosine kinase from the sponge *Geodia cydonium*. Cell Adhesion Commun 4:327–339

Pancer Z, Kruse M, Müller I, Müller WEG (1997) On the origin of adhesion receptors of Metazoa: cloning of the integrin alpha subunit cDNA from the sponge *Geodia cydonium*. Mol Biol Evol 14:391–398

Perovic S, Prokic I, Krasko A, Müller IM, Müller WEG (1999) Origin of neuronal-like receptors in Metazoa: cloning of a metabotropic glutamate/GABA-like receptor from the marine sponge *Geodia cydonium*. Cell Tissue Res 296:395–404

Pestka S, Langer JA, Zoon KC, Samuel CE (1987) Interferons and their actions. Annu Rev Biochem 56:727–777

Pfeifer K, Schröder HC, Rinkevich B, Uhlenbruck G, Hanisch FG, Kurelec B, Scholz P, Müller WEG (1992) Immunological and biological identification of tumor necrosis factor in sponges: role of this factor in the formation of necrosis in xenografts. Cytokine 4:161–169

Pfeifer K, Haasemann M, Gamulin V, Bretting H, Fahrenholz F, Müller WEG (1993) S-type lectins occur also in invertebrates: high conservation of the carbohydrate recognition domain in the lectin genes from the marine sponge *Geodia cydonium*. Glycobiology 3:179–184

Proksch P (1994) Defensive role for secondary metabolites from marine sponges and sponge-feeding nudibranchs. Toxicon 32:639–655

Radyuk SN, Klichko VI, Spinola B, Sohal RS, Orr WC (2001) The peroxiredoxin gene family in *Drosophila melanogaster*. Free Radic Biol Med 31:1090–1100

Rebouillat D, Hovnanian A, Marie I, Hovanessian AG (1999) The 100-kDa 2′,5′-oligoadenylate A synthetase catalyzing preferentially the synthesis of dimeric ppp2′pA5′ molecules is composed of three homologous domains. J Biol Chem 274:1557–1565

Richelle-Maurer E, Van de Vyver G (1999) Temporal and spatial expression of *EMH-3*, a homeobox-containing gene isolated from the freshwater sponge *Ephydatia muelleri*. Mech Ageing Dev 109:203–219

Samal B, Sun Y, Stearns G, Xie C, Suggs S, McNiece I (1994) Cloning and characterization of the cDNA encoding a novel human pre-B-cell colony-enhancing factor. Mol Cell Biol 14: 1431–1437

Sarma AS, Daum T, Müller WEG (1993) Secondary metabolites from marine sponges. Akademie gemeinnütziger Wissenschaften zu Erfurt, Ullstein-Mosby Verlag, Berlin

Schröder HC, Krasko A, Batel R, Skorokhod A, Pahler S, Kruse M, Müller IM, Müller WEG (2000) Stimulation of protein (collagen) synthesis in sponge cells by a cardiac myotrophin-related molecule from *Suberites domuncula*. FASEB J 14:2022–2031

Schütze J, Reis Custodio M, Efremova SM, Müller IM, Müller WEG (1999) Evolutionary relationship of Metazoa within the eukaryotes based on molecular data from Porifera. Proc R Soc Lond B 266:63–73

Schütze J, Krasko A, Diehl-Seifert B, Müller WEG (2001a) Cloning and expression of the putative aggregation factor from the marine sponge *Geodia cydonium*. J Cell Sci 114: 3189–3198

Schütze J, Skorokhod A, Pahler S, Müller IM, Müller WEG (2001b) Molecular evolution of metazoan extracellular matrix: cloning and expression of structural proteins from the demosponges *Suberites domuncula* and *Geodia cydonium*. J Mol Evol 53:402–415

Seimiya M, Ishiguro H, Miura K, Watanabe Y, Kurosawa Y (1994) Homeobox-containing genes in the most primitive Metazoa, the sponge. Eur J Biochem 221:219–225

Sen GC, Lengyel P (1992) The interferon system: a bird's eye view of its biochemistry. J Biol Chem 267:5017–5022

Sen S, Kundu G, Mekhail N, Castel J, Misono K, Healy B (1990) Myotrophin: purification of a novel peptide from spontaneously hypertensive rat heart that influences myocardial growth. J Biol Chem 265:16635–16643

Shalak V, Kaminska M, Mitnacht-Kraus R, Vandenabeele P, Clauss M, Mirande M (2001) The EMAPII cytokine is released from the mammalian multisynthetase complex after cleavage of its p43/proEMAPII component. J Biol Chem 276:23769–23776

Shiraishi T, Kuroiwa A, Shirakusa T, Kawahara K, Yoneda S, Kitano K, Okabayashi K, Iwasaki A (1997) Free radical-mediated tissue injury in acute lung allograft rejection and the effect of superoxide dismutase. Ann Thorac Surg 64:821–825

Sil P, Kandaswamy V, Sen S (1998) Increased protein kinase C activity in myotrophin-induced myocyte growth. Circ Res 82:1173–1188

Simpson TL (1984) The cell biology of sponges. Springer, Berlin Heidelberg New York

Smith WC, Harland RM (1992) Expression cloning of noggin, a new dorsalizing factor localized to the Spemann organizer in *Xenopus* embryos. Cell 70:829–840

Smith WC, Knecht AK, Wu M, Harland RM (1993) Secreted noggin protein mimics the Spemann organizer in dorsalizing mesoderm. Nature 361:547–549

Smith LC, Hildemann WH (1986) Allogeneic cell interactions during graft rejection in *Callyspongia diffusa* (Porifera; Demospongia); a study with monoclonal antibodies. Proc R Soc Lond B 226:465–477

Spemann H (1936) Experimentelle Beiträge zu einer Theorie der Entwicklung. Springer, Berlin Heidelberg New York

Spradling A, Drummond-Barbosa D, Kai T (2001) Stem cells find their niche. Nature 414:98–104

Utans U, Arceci RJ, Yamashita Y, Russell ME (1995) Cloning and characterization of allograft inflammatory factor-1: a novel macrophage factor identified in rat cardiac allografts with chronic rejection. J Clin Invest 95:2954–2962

Utans U, Quist WC, McManus BM, Wilson JE, Arceci RJ, Wallace AF, Russell ME (1996) Allograft inflammatory factory-1. A cytokine-responsive macrophage molecule expressed in transplanted human hearts. Transplantation 61:1387–1392

Valenzuela DM, Economides AN, Rojas E, Lamb TM, Nunez L, Jones P, Ip NY, Espinosa R, Brannan CI, Gilbert DJ, Copeland NG, Jenkins NA, LeBeau MM, Harland RM, Yancopoulos GD (1995) Identification of mammalian noggin and its expression in the adult nervous system. J Neurosci 15:6077–6084

Vogel S (1977) Current-induced flow through living sponges in nature. Proc Natl Acad Sci USA 74:2069–2071

Wagner-Hülsmann C, Bachinski N, Diehl-Seifert B, Blumbach B, Steffen R, Pancer Z, Müller WEG (1996) A galectin links the aggregation factor to cells in the sponge [*Geodia cydonium*] system. Glycobiology 6:785–793

Weinbaum G, Burger MM (1973) A two-component system for surface guided reassociation of animal cells. Nature 244:510–512

Wiens M, Kuusksalu A, Kelve M, Müller WEG (1999) Origin of the interferon-inducible (2′-5′)oligoadenylate synthetases: cloning of the (2′-5′)oligoadenylate synthetase from the marine sponge *Geodia cydonium*. FEBS Lett 462:12–18

Wiens M, Krasko A, Müller CI, Müller WEG (2000a) Molecular evolution of apoptotic pathways: cloning of key domains from sponges (Bcl-2 homology domains and death domains) and their phylogenetic relationships. J Mol Evol 50:520–531

Wiens M, Krasko A, Müller IM, Müller WEG (2000b) Increased expression of the potential proapoptotic molecule DD2 and increased synthesis of leukotriene B_4 during allograft rejection in a marine sponge. Cell Death Diff 7:461–469

Wiens M, Diehl-Seifert B, Müller WEG (2001) Sponge Bcl-2 homologous protein (BHP2-GC) confers selected stress resistance to human HEK-293 cells. Cell Death Diff 8:887–898

Wiens M, Mangoni A, D'Esposito M, Fattorusso E, Korchagina N, Schröder HC, Grebenjuk VA, Krasko A, Batel R, Müller IM, Müller WEG (2003) The molecular basis for the evolution of the metazoan bodyplan: extracellular matrix-mediated morphogenesis in marine demosponges. J Mol Evol (in press)

Wilson HV (1907) On some phenomena of coalescence and regeneration in sponges. J Exp Zool 5:245–258

Yang Y, Rao NS, Walker E, Sen S, Qin J (1997) Nuclear magnetic resonance assignment and secondary structure of an ankyrin-like repeat-bearing protein: myotrophin. Protein Sci 6:1347–1351

Yoshida S, Ono M, Shono T, Izumi H, Ishibashi T, Suzuki H, Kuwano M (1997) Involvement of interleukin-8, vascular endothelial growth factor, and basic fibroblast growth factor in tumor necrosis factor alpha-dependent angiogenesis. Mol Cell Biol 17:4015–4023

Zaheer A, Fink BD, Lim R (1993) Expression of glia maturation factor beta mRNA and protein in rat organs and cells. J Neurochem 60:914–920

Zahn RK, Müller WEG, Reinmüller J, Geisert M, Michaelis M, Pondeljak V, Beyer R (1976) Species-specific aggregation factor in sponges. Characterization of the large circular proteid particle. Cell Diff 5:129–137

Zhou A, Hassel BA, Silverman RH (1993) Expression cloning of 2-5A-dependent RNAase: an uniquely regulated mediator of interferon action. Cell 72:753–765

Functional Convergence of Invertebrate and Vertebrate Cytokine-Like Molecules Based on a Similar Lectin-Like Activity

A. Beschin,[1] M. Bilej,[2] S. Magez,[1] R. Lucas,[3] and P. De Baetselier[1]

Abstract. It is generally accepted that the action of cytokines results from their binding to specific receptors. However, many cytokines possess lectin-like activity that may be essential for the expression of their full biological activities. This review focuses on the physiological relevance of the lectin-like activity of cytokines during the innate immune response in mammals, using TNF as an illustrative example. Moreover, we will show that TNF displays functional analogies with a defense molecule from the earthworm *Eisenia foetida* termed CCF. These analogies are not reflected by primary sequence homology between CCF and TNF but are particularly based on a similar lectin-like activity/domain. Hence, from a phylogenetic point of view, the lectin-like activity/domain of CCF and TNF may represent an essential recognition mechanism that has been functionally conserved during the innate immune response of invertebrates and vertebrates as a result of convergent evolution.

1 Introduction

In vertebrates, cytokines are the major regulators of the host defense processes, and as such are involved in responses to exogenous and endogenous insults, repair and restoration of tissue homeostasis. Based on the assumption that invertebrates, as vertebrates, possess factors regulating responses to infection or wounding, studies dealing with the evolution of immunity have focussed on the isolation and characterization of putative cytokine-related molecules from invertebrates. Until recently, most of our knowledge of cytokine- and cytokine receptor-like molecules in invertebrates relied on functional assays. Typical experimental approaches to identify invertebrate cytokine-like or cytokine receptor-like molecules have evaluated the sensitivity of invertebrate immunocytes to vertebrate cytokine action, or vice versa the responsiveness of vertebrate immune cells to invertebrate factors. Moreover, antibodies neutralizing the activity of vertebrate cytokines were used to screen for cross-reactivity with invertebrate factors in hemolymph, immunocytes or neuroendocrine

[1]Department of Immunology, Parasitology and Ultrastructure, Flemish Interuniversity Institute for Biotechnology, Free University Brussels (VUB), Pleinlaan 2, 1050 Brussels, Belgium
[2]Department of Immunology, Institute of Microbiology, Academy of Sciences of the Czech Republic, Videnska 1083, 142 20 Prague 4, Czech Republic
[3]Biochemical Pharmacology, University of Konstanz, Universitatsstrasse 10, 78457 Konstanz, Germany

Progress in Molecular and Subcellular Biology
A. Beschin (Ed.) Invertebrate Cytokines

cells. In some studies, similarities at the physicochemical level (molecular weight, pI, . . .) of vertebrate cytokines and functional invertebrate analogues were considered. As such, a phylogenetic relationship between invertebrate cytokine-like molecules and vertebrate counterparts could not be convincingly demonstrated (for review Beschin et al. 2001). This paper presents recent evidence suggesting that mammalian cytokines contribute to innate immune function through lectin-like interactions. Furthermore, we summarize our own data on annelids, suggesting that the occurrence of invertebrate cytokine-like molecules and vertebrate factors does not reflect a common evolutionary origin but is rather a consequence of a convergence of function based on similar lectin-like activities. Finally, these data will be discussed in the current context of cytokine emergence during phylogeny.

2
Cytokine Activities Mediated by Lectin-Like Interactions

Functional pleiotropy and redundancy characterize mammalian cytokines. Each cytokine is usually produced by various cell types and acts on multiple target cells and tissues. It is generally accepted that the action of cytokines results from their binding to a membrane receptor. However, despite extensive studies revealing that many cytokines share receptor subunits, the signal transduction pathway is often cytokine- and target cell-specific. It was postulated that the specific association of a cytokine with other cell surface complexes could provide the specific signaling system (Fukushima and Yamashita 2001; Fukushima et al. 2001; Cebo et al. 2001, 2002a,b).

Many cytokines are bifunctional molecules having beside a receptor-binding domain, a carbohydrate recognition domain localized in opposite to the receptor-binding site (Lucas et al. 1994a; Zanetta et al. 1996a; Cebo et al. 2002b). This carbohydrate binding domain was suggested to be crucial for the expression of the full biological activities of cytokines. In this regard, IL-1α/β, IL-2 to IL-8, IL-10, IL-12, IFN-γ, TNF, TGF-β1, GM-CSF and FGF-β have been shown to interact with soluble glycosaminoglycans (GAG) or GAG-side chains of proteoglycans expressed on the cell surface of mammalian cells, thereby modulating their biological activities (Ramsden and Rider 1992; Grimaud and Lortat-Jacob 1994; Witt and Lander 1994; Clarke et al. 1995; Lortat-Jacob et al. 1994, 1997; Lyon et al. 1997; Lipscombe et al. 1998; Salek-Ardakani et al. 2000; Zanetta et al. 2002). For example, the binding of cytokines to cell surface proteoglycans or to the extracellular matrix proteoglycans, by providing a local tissue-bound reservoir of cytokines, may facilitate their interaction with high affinity receptors and/or trigger receptor dimerization and biological responses (Tanaka et al. 1998; Cebo et al. 2001). In turn, soluble GAG can modulate the activity of cytokines by competing with membrane receptors or by protecting the cytokines from proteolytic inactivation in the circulation (Daubener et al. 1995; Fernandez-Botran et al. 1999). More recently, it has been proposed that the carbohydrate-binding domains of some cytokines (IL-1α/β, IL-2, IL-4, IL-6) would be responsible for the association of the cytokine receptor with molecular complexes comprising specific kinase/phosphatase involved in cytokine-specific signal transduction (Zanetta et al. 1996b, 1998; Luton et al. 1997; Fukushima and Yamashita 2001; Cebo et al. 2002a,b). Accordingly, only cells expressing simultaneously the receptor and the carbohydrate ligand at their surface could respond optimally to a particular cytokine. Moreover, the cell-specific association of the cytokine receptor with a peculiar surface molecular complex could explain why cytokines sharing a particular receptor subunit but having distinct lectin-like activity might activate specifically certain cell types.

Beside interaction with mammalian cells, certain cytokines, in particular IL-1, IL-2 and TNF, have also been shown to directly interact with various pathogens through *N,N′*-diacetylchitobiose and Man5/6GlcNAc2-R lectin-like interactions (Luo et al. 1993; Lucas et al. 1994a,b; Olson et al. 1996; Zanetta et al. 1998; Magez et al. 2001; Table 1). IL-1 binds virulent *Escherichia coli*, acting as a growth factor for the bacteria. Similarly, IL-2 binds *Candida albicans* and acts as a growth factor for virulent *E. coli* and *Leishmamia mexicana*. In addition, TNF was shown to interact through lectin-like interactions with Gram-negative bacteria, with trypanosome parasites or with fungal β-1,3 glucans. It has been suggested that the lectin-like domains of cytokines represent pathogen-specific recognition sites that can contribute to the elimination of pathogens via opsonization and/or leukocyte activation (Luo et al. 1993; George 1994; Lucas et al. 1994a,b). However, other reports suggest that the binding of cytokines on cell wall carbohydrate moiety of pathogens may induce specific immunodeficiencies and interfere with the resolution of the disease (Zanetta et al. 1996b, 1998).

Table 1. Lectin properties of cytokines: specific recognition of microbes? (De Baetselier et al. 2002, reprinted with permission of the publisher)

	Carbohydrate ligands	Pathogen interaction
IL-1	*N,N′*-Diacetylchitobiose Man5/6GlcNAc2-R Man core of VSG	Growth factor for virulent *E. coli* Binding to *T. brucei* VSG
IL-2	*N,N′*-Diacetylchitobiose Man5/6GlcNAc2-R	Binding to *Candida albicans* Growth factor for virulent *E. coli* Growth factor for *Leishmania mexicana*
IL-6	β-1,3-Glucan	Opsonization of yeast
TNF	*N,N′*-Diacetylchitobiose Man5/6GlcNAc2-R β-1,3-Glucan	Opsonization of *Shigella*, *E. coli* and yeast Lysis of African and American trypanosomes
GM-CSF	Glycosaminoglycan	Growth factor for virulent *E. coli*
βFGF	Heparin sulfate	Growth factor for African trypanosomes

The following paragraphs will focus on the physiological relevance of the lectin-like activity of TNF during the innate immune response in mammals.

3
On the Physiological Relevance of the Lectin-Like Activity of TNF: Interaction with Microbes

The cytokine TNF was found to exhibit a direct trypanolytic activity on the bloodstream forms of the protozoan parasite *Trypanosoma brucei* by binding *N,N′*-diacetylchitobiose moieties on parasite surface components (Lucas et al. 1994a; Magez et al. 1997, 2001). Subsequent studies have shown that a similar interaction occurs between TNF and the protozoan parasite *T. cruzi* (Olivares Fontt et al. 1998). Since the trypanolytic activity of TNF on these parasites was recorded in vitro the question was addressed whether such activities are relevant in vivo. To tackle this important question, more extensive studies were carried out in the *T. brucei* model of infection, and the major observations and thereof-deduced conclusions are summarized below.

Infection with *T. brucei* was extensively documented to cause a profound activation of macrophages as manifested by the secretion of pro-inflammatory cytokines including TNF (Lucas et al. 1993; Magez et al. 1993). Subsequently, the major cell surface component of the parasite namely the Variant Surface Glycoprotein (VSG), a glycophosphatidylinositol GPI-anchored protein that forms an ever changing dense protective coat, was identified as the main TNF-inducing component and this activity was localized in the GPI-anchor of the molecule (Magez et al. 1998). Hereby, it should be emphasized that African trypanosomes have adopted a mechanism of antigenic variation of their VSG, and during oscillating waves of parasitemia large amounts of VSG molecules (estimated at 4×10^{15} molecules or an equivalent of 200 µg of protein) and hence of VSG-specific GPI moieties, are released in the circulation of the mammalian host. Consequently, there is a chronic VSG/GPI-mediated production of TNF in the host during infections with African trypanosomes. Similarly, American trypanosomes (i.e. *T. cruzi*) trigger via major GPI-anchored membrane components, such as mucin, host macrophages to produce TNF (Campos et al. 2001). Recent studies suggest that GPI and related structures from parasitic protozoa anchors engage members of the Toll-like receptor family to trigger the synthesis of inflammatory cytokines by macrophages (Almeida et al. 2000; Campos et al. 2001; Ropert et al. 2002). Moreover, GPI-anchored antigens presented by CD1 molecules were shown to induce the proliferation of NKT cells (Schofield et al. 1999; Procopio et al. 2002).

Based on the potential of TNF to exert a trypanolytic activity at least in vitro, the parasite-mediated induction of TNF (via GPI-linked molecules) could play a role in parasite control. This possibility was substantiated by analyzing *T. brucei* development either in mice treated with monoclonal antibodies specific for the lectin-like domain of TNF or in $TNF^{-/-}$ mice (Magez et al. 1997, 1999, 2002). In both experimental models (in situ neutralization of the lectin-like activity of TNF or total absence of TNF), a significant increase of the first wave of parasitemia was recorded, strongly suggesting that TNF contributes to parasite control at least during the early phase of infection. Hereby, it is appropriate to mention that whereas early bloodstream forms of *T. brucei* (i.e. before the ascent of the first peak of parasitemia) are resistant to TNF-mediated lysis in vitro, late bloodstream forms (i.e. at the peak of parasitemia) became fully TNF-sensitive. Furthermore, local rather than systemic TNF production may play a physiological role in parasite control. Indeed, peritoneal and liver macrophages from *T. brucei*-infected animals exert a trypanocidal activity ex vivo and such activity was totally absent in macrophages derived from $TNF^{-/-}$-infected mice (Magez et al. 2001). Collectively, these results, combined with a recent report showing that macrophage-derived TNF kills *T. gambiense* in mixed cell cultures (Dalouede et al. 2001), suggest that local production of TNF in inflamed tissues underlies an efficient killing mechanism in vivo that may account for early parasitemia control during African trypanosomosis.

Further studies aimed at the identification of the ligand for TNF binding and uptake by trypanosomes. Intriguingly, VSG was identified as the major ligand for TNF and the binding activity was localized in the N-linked carbohydrate side chain, more specifically the conserved *N,N′*-diacetylchitobiose-oligomannose (GlcNAc2-Man5–9) moiety of VSG (Magez et al. 2001). In fact, the VSG glycopeptide carrying the GlcNAc2-Man5–9 moiety potently inhibits TNF-mediated trypanosome killing by macrophages from *T. brucei*-infected animals. Hereby, it should be emphasized that the *N,N′*-diacetylchitobiose-oligomannose moiety of VSG is not exposed on the surface coat of the parasite but only accessible in the flagellar pocket where endocytosis takes place (Overath et al. 1997). Accordingly, TNF binding on *T. brucei* was restricted to the flagellar pocket and upon binding prompt endocytotic intracellular uptake of the mammalian cytokine took place (Magez et al. 1997). Subsequently, TNF-loaded endosomes fuse with lysosomes and the acidic pH of this organelle triggers TNF-mediated disruption of lysosomes, resulting in autolysis of the parasite.

To our knowledge, the accumulated results provide the first physiological link between a lectin-like cytokine domain, its carbohydrate ligand on a microorganism and an effector function – namely endocytosis-mediated lysis. Hereby, the parasite utilizes a single component, i.e. VSG, for the induction (via the GPI anchor) and uptake (via the N-linked carbohydrate chains) of this trypanocidal molecule (Fig. 1). Furthermore, from the conceptual point of view, at least in the field of parasite research, our results suggest how trypanosomes may circumvent the necessity to express large arrays of specific protein receptors for the uptake of host molecules required for regulated survival and development. Indeed, taking into account that trypanosomes have to cope with a

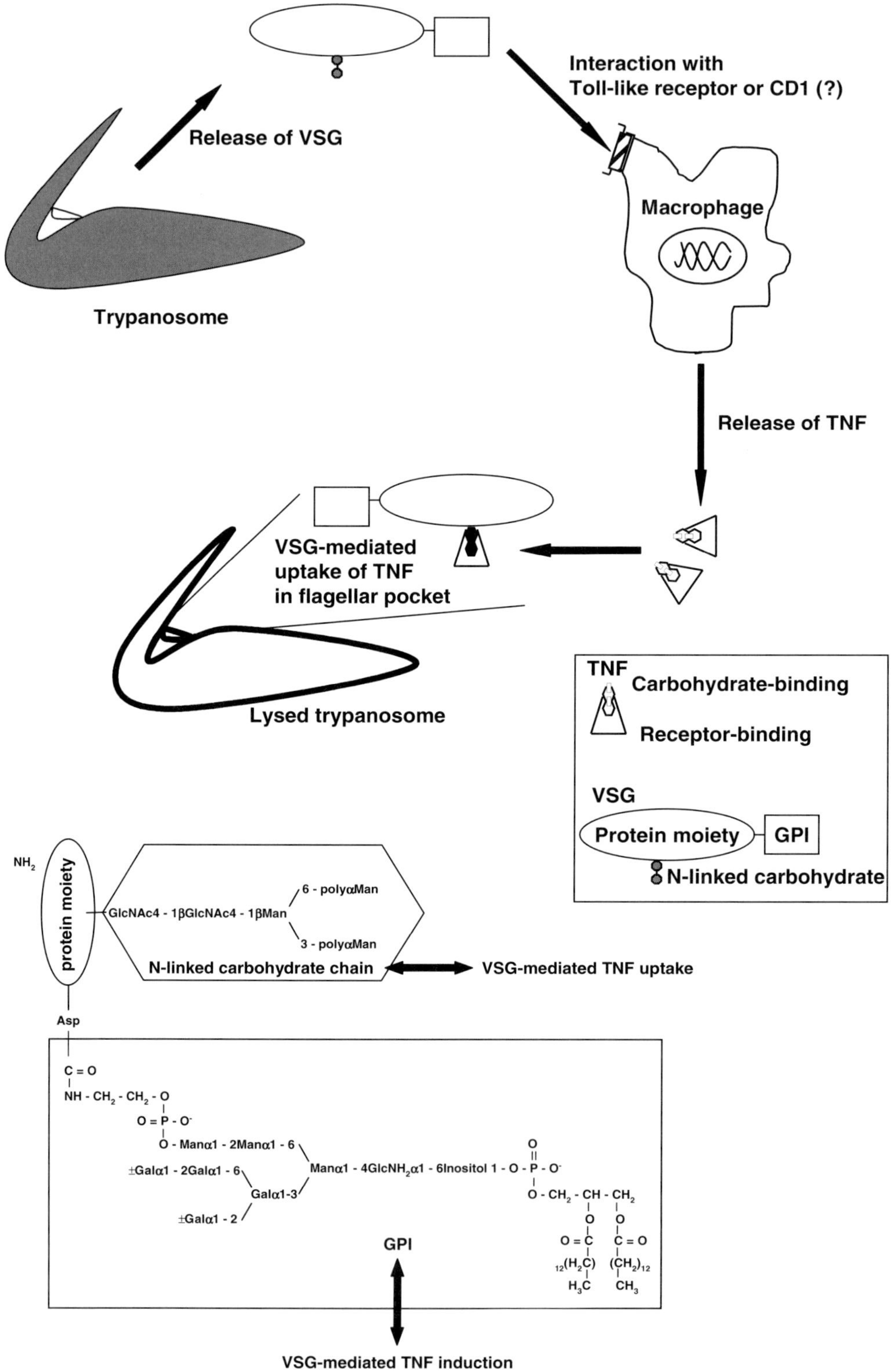
Interaction with
Toll-like receptor or CD1 (?)
Release of VSG
Macrophage
Trypanosome
Release of TNF
VSG-mediated
uptake of TNF
in flagellar pocket
Lysed trypanosome
TNF
Carbohydrate-binding
Receptor-binding
VSG
Protein moiety
GPI
N-linked carbohydrate
NH2
protein moiety
GlcNAc4 - 1βGlcNAc4 - 1βMan
6 - polyαMan
3 - polyαMan
N-linked carbohydrate chain
VSG-mediated TNF uptake
Asp
C = O
NH - CH2 - CH2 - O
O = P - O-
O - Manα1 - 2Manα1 - 6
±Galα1 - 2Galα1 - 6
Manα1 - 4GlcNH2α1 - 6Inositol 1 - O - P - O-
Galα1-3
±Galα1 - 2
O - CH2 - CH - CH2
O = C
C = O
12(H2C)
(CH2)12
H3C
CH3
GPI
VSG-mediated TNF induction

diversity of effector macromolecule analogues from different mammalian host species, the expression of few conserved but parasite-specific lectin ligands would offer a great advantage over the expression of multiple specific protein receptors. The same argument may hold true for other protozoan parasites and urges for a broadened view within parasite-receptor research.

4 On the Physiological Relevance of the Lectin-Like Activity of TNF: Interaction with Mammalian Cells

Our studies on the interaction between TNF and trypanosomes illustrate clearly the capacity of a mammalian cytokine to exert via its lectin-like domain a microbicidal activity. One may wonder whether this innate immune function, most probably shared with other cytokines displaying similar lectin-like domains (Table 1), is solely restricted to interactions with microbes or enables also physiologically relevant interactions with mammalian cells.

In fact, recent evidence suggests that the cytokine TNF contributes to inflammatory reactions in mammals through its lectin-like activity. Indeed, TNF was reported to increase the membrane conductance in endothelial cells, peritoneal macrophages and alveolar epithelial cells (Hribar et al. 1999; van der Goot et al. 1999). This effect does not rely on interaction with both types of TNF membrane receptors (p55 and p75) since it also occurred in cells isolated from p55/p75 TNF receptor-deficient mice. Hence, a TNF receptor-independent mechanism seemed to be implicated in this new physiological activity of TNF, and this ion channel gating activity was found to be mediated by the lectin-like domain of TNF. Furthermore, the TNF-mediated increased ion permeability could be inhibited by amiloride, an inhibitor of sodium transport, suggesting that the increase in whole cell current across the plasma membrane of mammalian cells resulted from the binding of TNF to an endogenous ion channel transporter or to a protein coupled to ion channels. The fact that the ion channel gating activity of TNF required an acidic environment is of interest (Hribar et al. 1999; van der Goot et al. 1999). This is quite reminiscent to the trypanolytic activity of TNF that also required an acidic pH in lysosomal compartments in order to cause autolysis of the parasite. Hence, it is possible

◄

Fig. 1. Dual role of African trypanosomes VSG during host–parasite interactions. At the peak stage of parasitemia, African trypanosomes release large amounts of VSG (from their entire surface) in the infected hosts. The released VSG activates host macrophages to produce the cytokine TNF. In turn, TNF interacts via its lectin-like activity/domain with the flagellar pocket exposed N-linked carbohydrate chain of VSG, leading to VSG-mediated endocytosis and subsequent TNF-mediated trypanolysis. The structure of the GPI anchor of VSG (Ferguson et al. 1985) is detailed. Asp is the C-terminal residue of the protein. (De Baetselier et al. 2002, reprinted with permission of the publisher)

that in lysosomes, TNF triggers changes in ion permeability resulting in lysosome disruption. Finally, the TNF-induced change in sodium permeability in type II alveolar epithelial cells was suggested to be important in edema resorption during acute inflammatory responses (Rezaiguia et al. 1997; Fukuda et al. 2001).

In conclusion, at least in the case of TNF, the lectin-like activity/domain of the cytokine may contribute to both innate defense mechanisms against microbes and inflammatory reactions in vertebrates (Fig. 2). Importantly, the ion gating activity of TNF, which possibly causes edema resorption, suggests that the lectin-like activity of TNF may contribute to the resolution of inflammatory processes during innate immune response in vertebrates (Hribar et al. 1999; van der Goot et al. 1999). In addition, the distinct receptor and carbohydrate recognition domain of TNF may trigger dual recognition/signaling pathways.

As summarized in the next paragraphs, we have identified an invertebrate defense molecule termed coelomic cytolytic factor (CCF) that represents a functional analogue of TNF. Interestingly, the analogies between CCF and TNF are solely based on a similar lectin-like activity/domain.

5 Coelomic Cytolytic Factor as an Earthworm Defense Molecule

CCF is a 42-kDa protein from the coelomic fluid of the earthworm *Eisenia foetida* (Bilej et al. 1995; Beschin et al. 1998). By efficiently binding the O-antigen of LPS, muramyl dipeptide of peptidoglycan, β-1,3-glucans and *N,N′*-diacetylchitobiose (β-1,4-*N*-acetylglucosidic link) (Beschin et al. 1998; Bilej et al. 2001), CCF triggers the activation of the prophenoloxidase cascade, an important defense mechanism both in protostomian and deuterostomian invertebrates. The broad saccharide recognition pattern of CCF results from the existence of distinct saccharide-binding domains. A first domain showing homology with the saccharide-binding motif of β-1,3-glucanases and invertebrate defense molecules located in the central part of CCF, is implicated in interactions with LPS and β-1,3-glucans. The C-terminal domain mediates interactions of CCF with *N,N′*-diacetylchitobiose and muramic acid (Fig. 3). These data provide evidence for the presence of spatially distinct lectin-like domains within an invertebrate defense molecule acting as a pattern recognition protein for cell wall components of yeast, Gram-negative and Gram-positive bacteria.

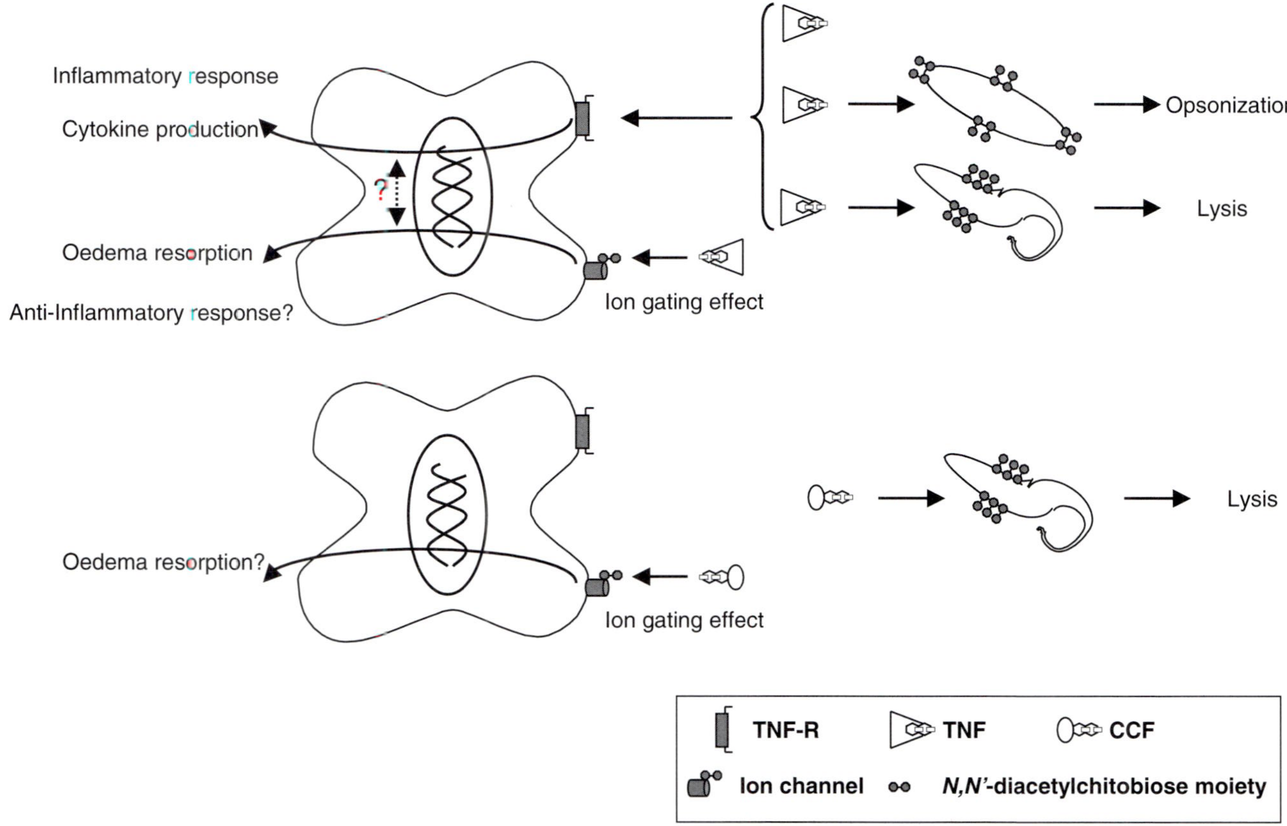

Fig. 2. Possible physiological relevance of the lectin-like activity of TNF. The interaction of TNF with microbes via the carbohydrate recognition domain resulting in the lysis of trypanosomes or by opsonizing Gram-negative bacteria may favor inflammatory immune response. Similar interaction of TNF with ion channels may result in anti-inflammatory immune response. Functional analogies of TNF with the CCF invertebrate defense molecule based on a similar *N,N'*-diacetylchitobiose lectin-like activity are also illustrated. Currently, it is not known whether (1) the signaling pathway triggered by TNF via TNF receptor or ion channel shares common mediators, (2) the ion gating effect of CCF causes edema resorption in mammalian cells, (3) CCF and TNF ion gating effects involve similar intracellular signaling

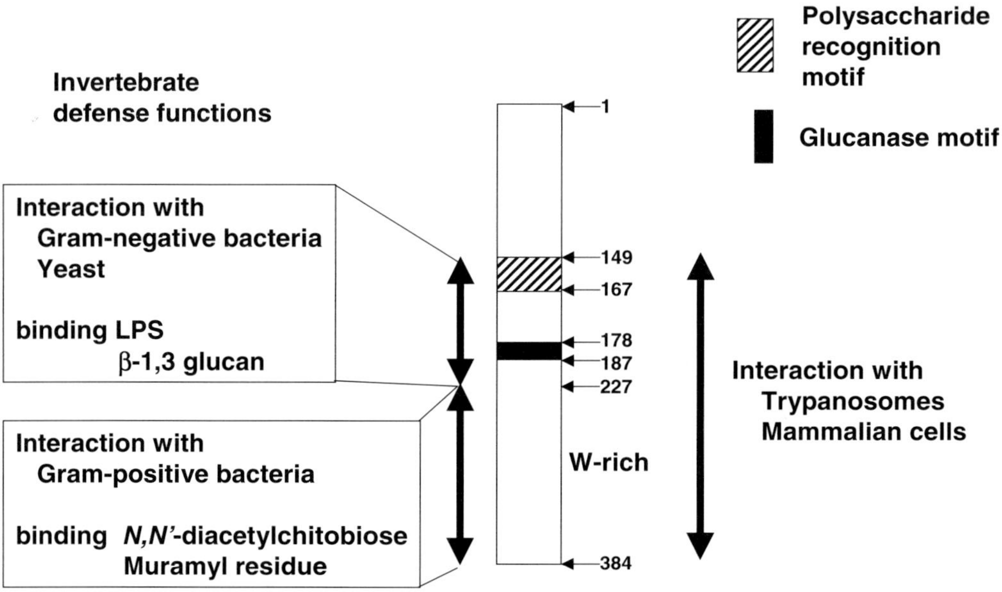

Fig. 3. Distinct functional pattern recognition domains in *Eisenia foetida* CCF. CCF encompasses two distinct domains interacting with pathogen-associated molecular patterns. The first domain located in the central part of CCF interacts with LPS and β-1,3-glucans while the C-terminal domain mediates interactions with *N,N'*-diacetylchitobiose and muramyl residues. Both domains are required for the interaction of CCF with trypanosomes and mammalian cells (Beschin et al., unpubl.)

6
Coelomic Cytolytic Factor as a TNF Analogue

Earthworms have provided for more than 30 years a useful model for comparative immunology. Their coelomic fluid exhibits a large variety of biological effects including bacteriostatic, hemolytic, proteolytic, and cytolytic activities that are involved in effective defense mechanisms against invaders (Bilej 1994a,b; Bilej et al. 2000). It was suggested that invertebrate molecules with tumorilytic activity might be analogous to the vertebrate cytokine TNF. CCF was originally identified in experiments aiming to characterize novel cytolytic factors from the coelomic fluid of *E. foetida*. We have observed that the coelomic fluid lyses TNF-sensitive tumor L929 cells in a protease-independent way, and subsequent isolation of the lytic proteins led to the identification of the 42-kDa protein CCF (Bilej et al. 1995; Beschin et al. 2002). In addition to the TNF-like lytic activity, CCF showed other similarities with this mammalian cytokine (Table 2). CCF is secreted by macrophage-like coelomocytes upon LPS stimulation while TNF is produced by macrophages

Table 2. Functional analogies of CCF with TNF

Lytic for L929 cells but in a TNF receptor-independent pathway
Increased production upon LPS or irradiation
Monoclonal antibodies cross-reactivity
Opsonizing properties
β-1,3-Glucan-binding activity
N,N′-Diacetylchitobiose-binding activity
Trypanolytic activity
Interaction with ion channels on mammalian cell surface

(Aggarwal 1985; Bilej et al. 1998). Moreover, CCF is involved in the opsonizing properties of the earthworm coelomic fluid (Bilej et al. 1995). Similarly, TNF was reported to provide an opsonin-like signal that mediates the attachment of bacteria to macrophages (Luo et al. 1993). Furthermore, CCF and TNF proteins bind β-1,3-glucans as well as *N,N′*-diacetylchitobiose via lectin-like interactions (Luo et al. 1993; Olson et al. 1996; Beschin et al. 1998, 1999). In addition, murine monoclonal antibodies elicited against the lectin-like domain of TNF (TIP domain), that is spatially distinct from the TNF receptor binding site (Lucas et al. 1994a), cross-react with CCF. Finally a monoclonal antibody elicited against CCF reacts with TNF (Bilej et al. 2000). Together these data suggest that CCF and TNF share similar β-1,3-glucan and *N,N′*-diacetylchitobiose lectin-like activities/domains.

As mentioned above (see Sect. 3), the lectin-like domain of TNF was shown to be involved in the killing of African and American trypanosomes by TNF. So, in view of the similar *N,N′*-diacetylchitobiose lectin-like activity of CCF and TNF, the possible trypanolytic activity of CCF was investigated (Beschin et al. 1999; Olivares Fontt et al. 2002). The coelomic fluid of *E. foetida* as well as purified CCF display potent trypanolytic activity that can be inhibited not only by anti-CCF monoclonal antibodies but also by *N,N′*-diacetylchitobiose and anti-TIP TNF antibodies. Vice versa, anti-CCF antibodies neutralize TNF-mediated trypanolysis. In addition, using the ability of CCF to trigger the prophenoloxidase cascade upon saccharide recognition in *E. foetida* coelomic fluid, the N-linked *N,N′*-diacetylchitobiose core of VSG, that acts as a protective coat on bloodstream forms of the African trypanosome *T. brucei*, was identified as a possible target for CCF on the trypanosome surface during trypanolytic events.

Finally, we have documented that CCF activates an amiloride-sensitive cationic channel in murine endothelial cells and macrophages via its *N,N′*-diacetylchitobiose lectin-like domain (Bloc et al. 2002). CCF treatment increases the outward current in macrophages from mice lacking both types of TNF-receptors excluding the interaction of the CCF with the TNF receptor. This activity of CCF and the underlying mechanism are reminiscent of the ion channel gating effect of TNF that was found to be mediated by the *N,N'*-diacetylchitobiose lectin-like domain of the molecule (see Sect. 4; Fig. 2).

Together, our data suggest that CCF and TNF share lectin-like activities that have been conserved as recognition mechanism in innate defense reactions in invertebrates and vertebrates, respectively. Importantly, despite their functional analogies, CCF and TNF do not show gene or amino acid sequence homology, indicating a lack of common evolutionary origin (Beschin et al. 1999).

7 Conclusions

Our comparative analysis shows that CCF and TNF display analogous functions but do not display phylogenetic relationships. As a consequence, CCF cannot be considered as an invertebrate cytokine homologue. In fact, the functional analogies between cytokines and their presumed invertebrate counterparts may actually not reflect homology, but rather result from molecular convergence (Conway Morris 2000) based on 3D-structural similarity of the lectin-like recognition domain. Thus, invertebrate defense molecules and mammalian cytokines may display a similar lectin-like activity that has been functionally conserved as recognition mechanism in innate immunity in invertebrates and vertebrates, respectively. These conclusions are corroborated by the works on protochordate IL-1-like molecules reviewed in the paper by Raftos et al. (this Vol.).

One of the challenges in comparative biology is to ascertain to what extent animals have homologous structures that derive from a common ancestor or whether any similarities are due to convergent evolution resulting from the need to perform similar biological functions. The striking superficial similarities between processes taking place in the invertebrates and vertebrates represent a great temptation to consider these two to be homologous, which means to have the same evolutionary origin. In this regard, the observed cross-reactivity of invertebrate cytokine-like molecules with antibodies elicited against mammalian cytokines only indicates that molecules from different phyla share epitopes/domains. However, homology must not be based on functional similarities but on entire gene or protein sequence similarities (Klein 1997). Moreover, characteristics shared by two taxa but absent in their common ancestor must be considered analogous but not homologous (Klein 1997).

Despite continuing efforts (Beck et al. 1989, 2000; Beck 1998), amino acid or gene sequences of the putative invertebrate cytokine or cytokine receptor molecules remain scarce. Moreover, the lack of molecular evidence despite completion of the *Caenorhabditis elegans* and *Drosophila* genome sequences (*C. elegans* Sequencing Consortium 1998; Adams et al. 2000) for the presence of full length cytokine receptor or cytokine homologues puts doubts on the existence of a true homologue of vertebrate cytokines and their receptors. Currently, homologues of TGF-β- and TGF-β receptor-like molecules have been cloned from invertebrates including protostomes and deuterostomes (Padgett et al. 1987; Wharton et al. 1991; Arora et al. 1994; Stenzel et al. 1994; Newfeld and Gelbart 1995; Ren et al. 1996; Sanchez-Salazar et al. 1996; Miya et al. 1997a,b; Gomez-Escobar et al. 1998; Ponce et al. 1999). TGF-β signaling pathway based on the activation of specific Smad proteins is also conserved in diverse metazoan phyla (Arora et al. 1995; Hahn et al. 1996; Savage et al. 1996; Massagué 1998; Patterson and Padgett 2000). However, the cloning of invertebrate TGF-β- and TGF-β receptor-like homologues did not reveal members of the TGF-β subfamily (Ponce et al. 1999) but members closer to the bone morphogenetic protein subfamily (Georgi et al. 1990; Estevez et al. 1993; Brummel et al. 1994; Nellen et al. 1994; Penton et al. 1994; Xie et al. 1994; Arora et al. 1995; Ren et al. 1996; Massagué 1998; Mochii et al. 1999; Rusten et al. 2002). On the other hand, genes and corresponding proteins expressing domains found in vertebrate cytokines have been recently identified in *Drosophila*. In this regard, *domeless*, a gene coding for a transmembrane protein (Dome) has been found to contain a cytokine-binding homology module with striking similarity to the IL-6 receptor family (Brown et al. 2001; Chen et al. 2002). By interacting with Unpaired, which lacks similarity with known proteins, Dome transduces a signal activating the invertebrate JAK/STAT pathway. Similarly, Eiger, a transmembrane protein with a C-terminal TNF homology domain – a 150 amino acid domain shared by all members of the TNF superfamily – and its ligand Wengen, a member of the TNF receptor superfamily – all sharing an homologous cysteine-rich domain – have been recently cloned in *Drosophila* (Igaki et al. 2002; Kanda et al. 2002; Moreno et al. 2002). Both molecules interact with each other through their TNF receptor and TNF homology domain, respectively, activating the JNK pathway. Interestingly, the signaling pathways of invertebrate TGF-β receptor-like molecules, Dome and Wengen – based on Smad, JAK/STAT and JNK, respectively – have been conserved through evolution (Savage et al. 1996; Caffrey et al. 1999; Luo and Dearolf 2001). The closer genetic relationship of invertebrate TGF-β superfamily to mammalian bone morphogenetic proteins, and the requirement of Dome, Eiger and Wengen in the regulation of the fly development (Ghiglione et al. 2002; Igaki et al. 2002; Kanda et al. 2002; Moreno et al. 2002), suggest that these molecules may not be involved in immune response in invertebrates. Importantly, the homology of Dome, Eiger and Wengen with vertebrate cytokines is restricted to a few critical amino acids that moreover belong to common protein domains that are not restricted to cytokines but mainly expressed by proteins involved in regulation of cell death and development (Igaki et al. 2002; Kanda et al. 2002). These observations support the postulate that the diversity of the proteome of vertebrates mainly results from the arrangement of domains eventually preexisting in invertebrates into a richer collection of domain architectures rather than from the presence of vertebrate-specific motif (Lander et al. 2001). In addition, the low sequence conservation and the homology restricted to common protein domains indicate that the so-far characterized invertebrate cytokine- or cytokine receptor-like molecules have gained diversity (pleiotropicity) and have been recruited by the immune system with the emergence of vertebrates.

Acknowledgements. This work was supported by the Foundation for Scientific Research-Flanders (G.0169.01), the Grant Agency of the Czech Republic (310/02/1437), and by Institutional Research Concept (AV0Z5020903) and was performed in the context of an Interuniversity Attraction Pole Program.

References

Adams MD, Celniker SE, Holt RA, Evans CA, Gocayne JD, Amanatides PG et al (2000) The genome sequence of *Drosophila melanogaster*. Science 287:2185–2195

Aggarwal BB (1985) Human tumor necrosis factor. Production, purification and characterization. J Biol Chem 260:2345–2354

Almeida IC, Camargo MM, Procopio DO, Silva LS, Mehlert A, Travassos LR, Gazzinelli RT, Ferguson MA (2000) Highly purified glycosylphosphatidylinositols from *Trypanosoma cruzi* are potent proinflammatory agents. EMBO J 19:1476–1485

Arora K, Levine MS, O'Connor MM (1994) The *screw* gene encode a ubiquitous expressed member of the TGF-beta family required for specification of dorsal cell fates in the *Drosophila* embryo. Genes Dev 8424:2588–2601

Arora K, Dai H, Kazuko SG, Jamal J, O'Connor MB, Letsou A, Warrior R (1995) The *Drosophila schnurri* gene acts in the Dpp/TGF-beta signaling pathway and encodes a transcription factor homologous to the human MBP family. Cell 81:781–790

Beck G (1998) Macrokines: invertebrate cytokine-like molecules. Front Biosci 3:559–569

Beck G, O'Brien RF, Habicht GS (1989) Invertebrate cytokines: the phylogenetic emergence of interleukin-1. BioEssays 11:62–67

Beck G, Ellis TW, Truong N (2000) Characterization of an IL-1 receptor from *Asterias forbesi* coelomocytes. Cell Immunol 203:66–73

Beschin A, Bilej M, Hanssens F, Raymakers J, van Dyck E, Revets H, Brys L, Gomez J, De Baetselier P, Timmermans M (1998) Identification and cloning of a glucan- and lipopolysaccharide-binding protein from *Eisenia foetida* earthworm involved in the activation of prophenoloxidase cascade. J Biol Chem 273:24948–24954

Beschin A, Bilej M, Brys L, Torreele E, Lucas R, Magez S, De Baetselier P (1999) Convergent evolution of cytokines. Nature 400:627–628

Beschin A, Bilej M, Torreele E, De Baetselier P (2001) On the existence of cytokines in invertebrates. Cell Mol Life Sci 58:801–814

Beschin A, De Baetselier P, Bilej M (2002) CCF, an invertebrate analogue of TNF, is not related to the other lytic components from *Eisenia foetida* earthworm. BioEssays 24:974

Bilej M (1994a) Cellular defense mechanisms. In: Vetvicka V, Sima P, Cooper EL, Bilej M, Roch P (eds) Immunology of annelids. CRC Press, Boca Raton, pp 167–200

Bilej M (1994b) Humoral defense mechanisms. In: Vetvicka V, Sima P, Cooper EL, Bilej M, Roch P (eds) Immunology of annelids. CRC Press, Boca Raton, pp 245–162

Bilej M, Brys L, Beschin A, Lucas R, Vercauteren E, Hanusova R, De Baetselier P (1995) Identification of a cytolytic protein in the coelomic fluid of *Eisenia foetida* earthworms. Immunol Lett 45:123–128

Bilej M, Rossmann P, Sinkora M, Hanusova R, Beschin A, Raes G, De Baetselier P (1998) Cellular expression of the cytolytic factor in earthworms *Eisenia foetida*. Immunol Lett 60:23–29

Bilej M, De Baetselier P, Beschin A (2000) Antimicrobial defense of the earthworm. Folia Microbiol 45:283–300

Bilej M, De Baetselier P, van Dijck E, Stijlemans B, Colige A, Beschin A (2001) Distinct carbohydrate recognition domains of an invertebrate defense molecule recognize Gram-negative and Gram-positive bacteria. J Biol Chem 276:45840–45847

Bloc A, Lucas R, Van Dijck E, Bilej M, Dunant Y, De Baetselier P, Beschin A (2002) An invertebrate defense molecule activates membrane conductance in mammalian cells by means of its lectin-like domain. Dev Comp Immunol 26:35–43

Brown S, Hu N, Hombria JC (2001) Identification of the first invertebrate interleukin JAK/STAT receptor, the *Drosophila* gene *domeless*. Curr Biol 11:1700–1705

Brummel TJ, Twombly V, Marques G, Wrana JL, Newfeld SJ, Attisano L, Massague J, O'Connor MB, Gelbart WM (1994) Characterization and relationship of Dpp receptors encoded by the saxophone and thick veins genes in *Drosophila*. Cell 78:251–261

Caffrey DR, O'Neill LA, Shields DC (1999) The evolution of the MAP kinase pathways: coduplication of interacting proteins leads to new signaling cascades. J Mol Evol 49:567–582

Campos MA, Almeida IC, Takeuchi O, Akira S, Valente EP, Procopio DO, Travassos LR, Smith JA, Golenbock DT, Gazzinelli RT (2001) Activation of Toll-like receptor-2 by glycosylphosphatidylinositol anchors from a protozoan parasite. J Immunol 167:416–423

Cebo C, Dambrouck T, Maes E, Laden C, Strecker G, Michalski JC, Zanetta JP (2001) Recombinant human interleukins IL-1alpha, IL-1beta, IL-4, IL-6, and IL-7 show different and specific calcium-independent carbohydrate-binding properties. J Biol Chem 276:5685–5691

Cebo C, Durier V, Lagant P, Maes E, Florea D, Lefebvre T, Strecker G, Vergoten G, Zanetta JP (2002a) Function and molecular modeling of the interaction between human interleukin 6 and its HNK-1 oligosaccharide ligands. J Biol Chem 277:12246–12252

Cebo C, Vergoten G, Zanetta JP (2002b) Lectin activities of cytokines: functions and putative carbohydrate-recognition domains. Biochim Biophys Acta 1572:422–434

Chen HW, Chen X, Oh SW, Marinissen MJ, Gutkind JS, Hou SX (2002) *mom* identifies a receptor for the *Drosophila* JAK/STAT signal transduction pathway and encodes a protein distantly related to the mammalian cytokine receptor family. Genes Dev 16:388–398

Clarke D, Katoh O, Gibbs RV, Griffiths SD, Gordon MY (1995) Interaction of interleukin 7 (IL-7) with glycosaminoglycans and its biological relevance. Cytokine 7:325–330

Conway Morris S (2000) Evolution: bringing molecules into the fold. Cell 100:1–11

C. elegans Sequencing Consortium (1998) Genome sequence of the nematode *C. elegans*: a platform for investigating biology. Science 282:2012–2018

Dalouede S, Bouteille B, Moynet D, De Baetselier P, Courtois P, Lemesre JL, Buguet A, Cespuglio R, Vincendeau P (2001) Human macrophage tumor necrosis factor (TNF)-alpha production induced by *Trypanosoma brucei gambiense* and the role of TNF-alpha in parasite control. J Infect Dis 183:988–991

Daubener W, Nockemann S, Gutsche M, Hadding U (1995) Heparin inhibits the antiparasitic and immune modulatory effects of human recombinant interferon-gamma. Eur J Immunol 25: 688–692

De Baetselier P, Beschin A, Lucas R, Bilej M, Stijlemans B, Magez S (2002) The functional relevance of the lectin-like activities of cytokines: TNF as an illustrative example. In: Cooper EL, Beschin A, Bilej M (eds) A new model for analyzing antimicrobial peptides with biomedical applications. IOS Press, Amsterdam, pp 157–163

Estevez M, Attisano L, Wrana JL, Albert PS, Massague J, Riddle DL (1993) The *daf-4* gene encodes a bone morphogenetic protein receptor controlling *C. elegans dauer* larva development. Nature 365:644–649

Ferguson MA, Low MG, Cross GA (1985) Glycosyl-sn-1,2-dimyristylphosphatidylinositol is covalently linked to Trypanosoma brucei variant surface glycoprotein. J Biol Chem 260: 14547–14555

Fernandez-Botran R, Yan J, Justus DE (1999) Binding of interferon gamma by glycosaminoglycans: a strategy for localization and/or inhibition of its activity. Cytokine 11:313–325

Fukuda N, Jayr C, Lazrak A, Wang Y, Lucas R, Matalon S, Matthay MA (2001) Mechanisms of TNF-alpha stimulation of amiloride-sensitive sodium transport across alveolar epithelium. Am J Physiol Lung Cell Mol Physiol 280:L1258–L1265

Fukushima K, Yamashita K (2001) Interleukin-2 carbohydrate recognition modulates CTLL-2 cell proliferation. J Biol Chem 276:7351–7356

Fukushima K, Hara-Kuge S, Ideo H, Yamashita K (2001) Carbohydrate recognition site of interleukin-2 in relation to cell proliferation. J Biol Chem 276:31202–31208

George AJ (1994) Surface-bound cytokines – a possible effector-mechanism in bacterial immunity. Immunol Today 15:88–89

Georgi LL, Albert PS, Riddle DL (1990) *daf-1*, a *C. elegans* gene controlling dauer larva development, encodes a novel receptor protein kinase. Cell 61:635–645

Ghiglione C, Devergne O, Georgenthum E, Carballes F, Medioni C, Cerezo D, Noselli S (2002) The *Drosophila* cytokine receptor *domeless* controls border cell migration and epithelial polarization during oogenesis. Development 129:5437–5447

Gomez-Escobar N, Lewis E, Maizels RM (1998) A novel member of the transforming growth factor-beta (TGF-β) superfamily from the filarial nematodes *Brugia malayi* and *B. pahangi*. Exp Parasitol 88:200–209

Grimaud JA, Lortat-Jacob H (1994) Matrix receptors to cytokines: from concept to control of tissue fibrosis dynamics. Pathol Res Pract 190:883–890

Hahn SA, Schutte M, Hoque AT, Moskaluk CA, Da Costa LT, Rozenblum E, Weinstein CL, Fischer A, Yeo CJ, Hruban RH, Kern SE (1996) *DPC4*, a candidate tumor suppressor gene at human chromosome 18q21.1. Science 271:350–353

Hribar M, Bloc A, van der Goot FG, Fransen L, De Baetselier P, Grau GE, Bluethmann H, Matthay MA, Dunant Y, Pugin J, Lucas R (1999) The lectin-like domain of tumor necrosis factor-alpha increases membrane conductance in microvascular endothelial cells and peritoneal macrophages. Eur J Immunol 29:3105–3111

Igaki T, Kanda H, Yamamoto-Goto Y, Kanuka H, Kuranaga E, Aigaki T, Miura M (2002) Eiger, a TNF superfamily ligand that triggers the *Drosophila* JNK pathway. EMBO J 21:3009–3018

Kanda H, Igaki T, Kanuka H, Yagi T, Miura M (2002) Wengen, a member of the *Drosophila* tumor necrosis factor receptor superfamily, is required for eiger signaling. J Biol Chem 277: 28372–2835

Klein J (1997) Homology between immune response in vertebrates and invertebrates: does it exist? Scand J Immunol 46:558–564

Lander ES, Linton LM, Birren B, Nusbaum C, Zody MC, Baldwin J et al (2001) Initial sequencing and analysis of the human genome. Nature 409:860–921

Lipscombe RJ, Nakhoul AM, Sanderson CJ, Coombe DR (1998) Interleukin-5 binds to heparin/heparan sulfate. A model for an interaction with extracellular matrix. J Leukoc Biol 63:342–350

Lortat-Jacob H, Esterre P, Grimaud JA (1994) Interferon-gamma, an anti-fibrogenic cytokine which binds to heparan sulfate. Pathol Res Pract 190:920–922

Lortat-Jacob H, Garrone P, Banchereau J, Grimaud JA (1997) Human interleukin 4 is a glycosaminoglycan-binding protein. Cytokine 9:101–105

Lucas R, Magez S, de Leys R, Fransen L, Scheerlinck JP, Rampelberg M, Sablon E, De Baetselier P (1994a) Mapping the lectin-like activity of tumor necrosis factor. Science 263:814–817

Lucas R, Magez S, de Leys R, De Baetselier P (1994b) Cytokines: multifunctional pathogen-specific mammalian lectin? In: Van Driessche E, Fisher J, Beeckmans S, Bog-Hansen TC (eds) Lectins: biology, biochemistry, clinical biochemistry, vol 10. Textop, Hellerup, Denmark, pp 244–249

Lucas R, Magez S, Songa B, Darji A, Hamers R, De Baetselier P (1993) A role for TNF during African trypanosomiasis: involvement in parasite control, immunosuppression and pathology. Res Immunol 144:370–376

Luo G, Niesel DW, Shaban RA, Grimm EA, Klimpel GR (1993) Tumor necrosis factor alpha binding to bacteria: evidence for a high-affinity receptor and alteration of bacterial virulence properties. Infect Immun 61:830–835

Luo H, Dearolf CR (2001) The JAK/STAT pathway and *Drosophila* development. BioEssays 23: 1138–1147

Luton F, Legendre V, Gorvel JP, Schmitt-Verhulst AM, Boyer C (1997) Tyrosine and serine protein kinase activities associated with ligand-induced internalized TCR/CD3 complexes. J Immunol 158:3140–3147

Lyon M, Rushton G, Gallagher JT (1997) The interaction of the transforming growth factor-betas with heparin/heparan sulfate is isoform-specific. J Biol Chem 272:18000–18006

Magez S, Lucas R, Darji A, Songa EB, Hamers R, De Baetselier P (1993) Murine tumour necrosis factor plays a protective role during the initial phase of the experimental infection with *Trypanosoma brucei brucei*. Parasite Immunol 15:635–641

Magez S, Geuskens M, Beschin A, del Favero H, Verschueren H, Lucas R, Pays E, De Baetselier P (1997) Specific uptake of tumor necrosis factor-alpha is involved in growth control of *Trypanosoma brucei*. J Cell Biol 137:715–727

Magez S, Stijlemans B, Radwanska M, Pays E, Ferguson MA, De Baetselier P (1998) The glycosyl-inositol-phosphate and dimyristoylglycerol moieties of the glycosylphosphatidylinositol anchor of the trypanosome variant-specific surface glycoprotein are distinct macrophage-activating factors. J Immunol 160:1949–1956

Magez S, Radwanska M, Beschin A, Sekikawa K, De Baetselier P (1999) Tumor necrosis factor alpha is a key mediator in the regulation of experimental *Trypanosoma brucei* infections. Infect Immun 67:3128–3132

Magez S, Radwanska M, Stijlemans B, Xong HV, Pays E, De Baetselier P (2001) A conserved flagellar pocket exposed high mannose moiety is used by African trypanosomes as a host cytokine binding molecule. J Biol Chem 276:33458–33464

Magez S, Stijlemans B, Caljon G, Eugster HP, De Baetselier P (2002) Control of experimental *Trypanosoma brucei* infections occurs independently of lymphotoxin-alpha induction. Infect Immun70:1342–1351

Massagué J (1998) TGF-beta signal transduction. Annu Rev Immunol 67:753–791

Miya T, Morita K, Ueno N, Satoh N (1997a) An ascidian homologue of vertebrate Bmps-5–8 is expressed in midline of the anterior neuroectoderm and in the midline of the ventral epidermis of the embryo. Mech Dev 57:181–190

Miya T, Morita K, Suzuki A, Ueno N, Satoh N (1997b) Functional analysis of an ascidian homologue of vertebrate Bmp-2/Bmp-4 suggests its role in the inhibition of neural fate specification. Development 124:5149–5159

Mochii M, Yoshida S, Morita K, Kohara Y, Ueno N (1999) Identification of transforming growth factor-beta- regulated genes in *Caenorhabditis elegans* by differential hybridization of arrayed cDNAs. Proc Natl Acad Sci USA 96:15020–15025

Moreno E, Yan M, Basler K (2002) Evolution of TNF signaling mechanisms. JNK-dependent apoptosis triggered by eiger, the *Drosophila* homolog of the TNF superfamily. Curr Biol 12:1263

Nellen D, Affolter M, Basler K (1994) Receptor serine/threonine kinases implicated in the control of *Drosophila* body pattern by decapentaplegic. Cell 78:225–237

Newfeld SJ, Gelbart WM (1995) Identification of two *Drosophila* TGF-beta family members in the grasshopper *Schistocerca Americana*. J Mol Evol 41:155–160

Olivares Fontt E, De Baetselier P, Heirman C, Thielemans K, Lucas R, Vray B (1998) Effects of granulocyte-macrophage colony-stimulating factor and tumor necrosis factor alpha on *Trypanosoma cruzi* trypomastigotes. Infect Immun 66:2722–2727

Olivares Fontt E, Beschin A, van Dijck E, Vercruysse V, Bilej M, Lucas R, De Baetselier P, Vray B (2002) *Trypanosoma cruzi* is lysed by coelomic cytolytic factor-1, an invertebrate analogue of tumor necrosis factor, and induces phenoloxidase activity in the coelomic fluid of *Eisenia foetida foetida*. Dev Comp Immunol 26:27–34

Olson EJ, Standing JE, Griego-Harper N, Hoffman OA, Limper AH (1996) Fungal beta-glucan interacts with vitronectin and stimulates tumor necrosis factor alpha release from macrophages. Infect Immun 64:3548–3554

Overath P, Stierhof YD, Wiese M (1997) Endocytosis and secretion in trypanosomatid parasites: tumultuous traffic in a pocket. Trends Cell Biol 7:27–33

Padgett RW, St Johnston D, Gelbart WM (1987) A transcript from a *Drosophila* pattern gene predicts a protein homologous to the transforming growth factor-beta family. Nature 325: 81–84

Patterson GI, Padgett RW (2000) TGF-beta-related pathways: role in *Caenorhabditis elegans* development. Trends Genet 16:27–33

Penton A, Chen Y, Staehling-Hampton K, Wrana JL, Attisano L, Szidonya J, Cassill JA, Massague J, Hoffmann FM (1994) Identification of two bone morphogenetic protein type I receptors in *Drosophila* and evidence that Brk25D is a decapentaplegic receptor. Cell 78:239–250

Ponce MR, Micol JL, Peterson KJ, Davidson EH (1999) Molecular characterization and phylogenetic analysis of SpBPM5-7, a new member of the TGF-beta superfamily expressed in sea urchin embryos. Mol Biol Evol 16:634–645

Procopio DO, Almeida IC, Torrecilhas AC, Cardoso JE, Teyton L, Travassos LR, Bendelac A, Gazzinelli RT (2002) Glycosylphosphatidylinositol-anchored mucin-like glycoproteins from *Trypanosoma cruzi* bind to CD1d but do not elicit dominant innate or adaptive immune responses via the CD1d/NKT cell pathway. J Immunol 169:3926–3933

Ramsden L, Rider CC (1992) Selective and differential binding of interleukin (IL)-1 alpha, IL-1 beta, IL-2 and IL-6 to glycosaminoglycans. Eur J Immunol 22:3027–3031

Ren P, Lim CS, Johnsen R, Albert PS, Pilgrim D, Riddle DL (1996) Control of *C. elegans* larval development by neuronal expression of a TGF-beta homolog. Science 274:1389–1391

Rezaiguia S, Garat C, Delclaux C, Meignan M, Fleury J, Legrand P, Matthay MA, Jayr C (1997) Acute bacterial pneumonia in rats increases alveolar epithelial fluid clearance by a tumor necrosis factor-alpha-dependent mechanism. J Clin Invest 99:325–335

Ropert C, Ferreira LR, Campos MA, Procopio DO, Travassos LR, Ferguson MA, Reis LF, Teixeira MM, Almeida IC, Gazzinelli RT (2002) Macrophage signaling by glycosylphosphatidylinositol-anchored mucin-like glycoproteins derived from *Trypanosoma cruzi* trypomastigotes. Microbes Infect 4:1015–1025

Rusten TE, Cantera R, Kafatos FC, Barrio R (2002) The role of TGF beta signaling in the formation of the dorsal nervous system is conserved between *Drosophila* and chordates. Development 129:3575–3584

Salek-Ardakani S, Arrand JR, Shaw D, Mackett M (2000) Heparin and heparan sulfate bind interleukin-10 and modulate its activity. Blood 96:1879–1888

Sanchez-Salazar J, Pletcheer MT, Bennett RL, Brown SJ, Dandamudi TJ, Denell RE, Doctor JS (1996) The *Tribolium decapentaplegic* gene is similar in sequence, structure, and expression to the *Drosophila dpp* gene. Dev Genes Evol 206:237–246

Savage C, Das P, Finelli AL, Townsend SR, Sun CY, Baird SE, Padgett RW (1996) *Caenorhabditis elegans* genes *sma-2*, *sma-3*, and *sma-4* define a conserved family of transforming growth factor beta pathway components. Proc Natl Acad Sci USA 93:790–794

Schofield L, McConville MJ, Hansen D, Campbell AS, Fraser-Reid B, Grusby MJ, Tachado SD (1999) CD1d-restricted immunoglobulin G formation to GPI-anchored antigens mediated by NKT cells. Science 283:225–229

Stenzel P, Angerer LM, Smith BJ, Angerer RC, Vale WW (1994) The *univin* gene encodes a member of the transforming growth factor-beta superfamily with restricted expression in the sea urchin embryo. Dev Biol 1664:149–158

Tanaka Y, Kimata K, Adams DH, Eto S (1998) Modulation of cytokine function by heparan sulfate proteoglycans: sophisticated models for the regulation of cellular responses to cytokines. Proc Assoc Am Physicians110:118–125

Tandai-Hiruma M, Endo T, Kobata A (1999) Detection of novel carbohydrate binding activity of interleukin-1. J Biol Chem 274:4459–4466

Van der Goot FG, Pugin J, Hribar M, Fransen L, Dunant Y, De Baetselier P, Bloc A, Lucas R (1999) Membrane interaction of TNF is not sufficient to trigger increase in membrane conductance in mammalian cells. FEBS Lett 460:107–111

Wharton KA, Thomsen GH, Gelbart WM (1991) *Drosophila 60A* gene, another transforming growth factor-beta family member, is closely related to human bone morphogenetic proteins. Proc Natl Acad Sci USA 88 9214–9218

Witt DP, Lander AD (1994) Differential binding of chemokines to glycosaminoglycan subpopulations. Curr Biol 4:394–400
Xie T, Finelli AL, Padgett RW (1994) The *Drosophila saxophone* gene: a serine-threonine kinase receptor of the TGF-beta superfamily. Science 263:1756–1759
Zanetta JP, Scior T, Wantyghem J, Wermuth C, Aubery M, Strecker G, Michalski JC (1996a) Lectin activities of cytokines and growth factors: function and implications for pathology. Histol Histopathol 11:1101–1108
Zanetta JP, Alonso C, Michalski JC (1996b) Interleukin 2 is a lectin that associates its receptor with the T-cell receptor complex. Biochem J 318:49–53
Zanetta JP, Bonaly R, Maschke S, Strecker G, Michalski JC (1998) Differential binding of lectins IL-2 and CSL to *Candida albicans* and cancer cells. Glycobiology 8:221–225
Zanetta JP, Bindeus R, Normand G, Durier V, Lagant P, Maes E, Vergoten G (2002) Evidence for a lectin activity for human interleukin 3 and modeling of its carbohydrate recognition domain. J Biol Chem 277:38764–38771

Tunicate Cytokine-like Molecules and Their Involvement in Host Defense Responses

D. Raftos and S. Nair

Abstract. Tunicates (ascidians or sea squirts) are a large group of invertebrate chordates that are closely related to vertebrates. Their critical phylogenetic position has stimulated substantial interest in their host defense ("immune") responses. Whilst this interest has generated a wealth of knowledge regarding the humoral and cellular mechanisms that undertake defensive responses, there is less known about the regulation of those reactions. This chapter focuses on three cellular responses (cell proliferation, phagocytosis and chemotaxis) that are known to be regulated by cytophilic humoral molecules. Some of the humoral factors that affect these responses have functional and physicochemical similarities to vertebrate cytokines, like interleukin-1. However, the only regulatory molecules that have been characterized at a molecular level bear far greater similarity to C-type lectins or complement components.

1 Tunicates

1.1 General Characteristics

Tunicates (sea squirts) are a large group of invertebrates that frequently go unnoticed because of their sessile life styles (Berrill 1950). The earliest general description of this group comes from Aristotle, who likened them to the traditional leather wine gourds – askidia – that are still used throughout the Mediterranean. Hence, these animals are often referred to as ascidians, particularly in Europe and Japan. Tunicates are characterized by an outer covering, or tunic, that often forms a thick epidermal protective layer. The tunic is made from a matrix of protein and carbohydrate that, in the past, was thought to be analogous to plant cellulose.

The tunic provides tunicates not only with their name, which was established by Lamarck in 1816 (Berrill 1950), but also with their often unflattering external appearance (Fig. 1). The perplexing nature of many tunicate species has even led to discussions in the popular press as to whether they represent animal, vegetable or mineral. Their firm footing in the animal kingdom was put most eloquently by Dr. Archibald Grubb ... "*... I hasten to state that my observations of and on tunicates (sic) have been frequent, long, and even lurid,*

Department of Biological Sciences, Macquarie University, North Ryde, NSW, 2109, Australia

Progress in Molecular and Subcellular Biology
A. Beschin (Ed.) Invertebrate Cytokines

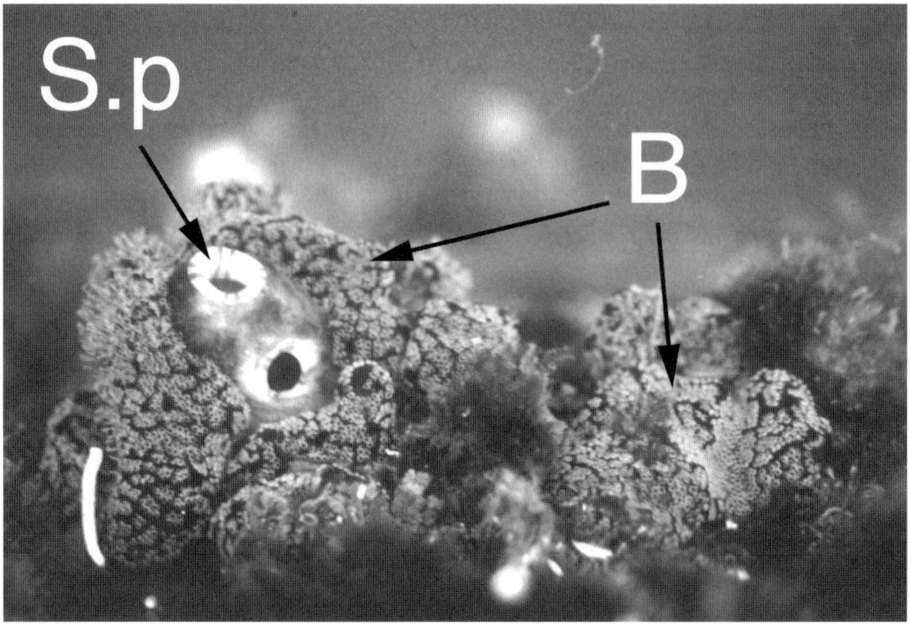

Fig. 1. Two species of tunicate from Birkenhead Point in Sydney Harbor, Australia. *S.p* A specimen of the solitary tunicate species, *Styela plicata*, which can be identified by its two white siphons. *B* A colonial tunicate, probably from the genus *Botrylloides*. (Photograph courtesy of Jane Radford and Jenny Norman, Department of Biological Sciences, Macquarie University)

and their facility to deprive me of hooks and sinkers has always suggested that they are animals, and a very lowly and objectionable animals at that. . . . Yes, if the tunicate (sic) isn't a lower form of animal life, the barnacle is a melon-plant, or I'm an onion" (from The Sydney Sun, 15 May 1945; Dakin 1980).

1.2 Tunicate Phylogeny

Whilst Dr. Grubb was correct that tunicates are animals, his belief that they are "*very lowly and objectionable animals*" was not. Rather, tunicates represent one of the most closely related taxa to the vertebrates. This relationship was first proposed in the 19th century by Kowalevsky (Berrill 1955). It was originally based on the identification among tunicate larvae of structures that appear to be homologous to phylotypic chordate characteristics, such as a postanal tail, dorsal nerve cord, notochord and pharyngeal gill slits (Garstang 1928). These structures are lost in adult tunicates because larvae undergo a radical metamorphosis. However, the larval homologies identified by Kowalevsky remain fundamentally important because they provide definitive clues to the origin of vertebrates (Berrill 1955; Dakin 1980). It is now widely believed that vertebrates evolved by the neotenic or paedomorphic development of tunicate-like larvae from an ancient protochordate stock (Garstang 1928; Berrill 1955). This close evolutionary relationship between tunicates and "higher" chordates has been confirmed by molecular phylogenetics, notably those employing 18S rRNA (Field et al. 1988).

1.3 The Tunicate Immune System

The phylogenetic proximity of tunicates to vertebrates has led to significant interest in their immunological capacities – generated by the belief that tunicates may deploy an evolutionary precursor of the vertebrate immune system. However, extensive analysis has shown that, in terms of anti-pathogen reactions, traits such as specificity, recognitive diversity and immunological memory are associated only with the adaptive immune responses of jawed vertebrates (gnathostomes) (Raftos and Raison 1992; Medzhitov and Janeway 1997). Intensive efforts to identify Ig-based elements of the adaptive immune system in invertebrates have provided no evidence for defensive immunoglobulins in any group other than gnathostomes.

Despite the absence of adaptive defense responses mediated by antibodies in tunicates, these animals do undertake "innate" immune responses that are usually mediated by components of the hemolymph (circulating body fluid).

1.4 Tunicate Hemolymph

1.4.1 Serum Component

The humoral component of tunicate hemolymph (designated "serum" in this chapter) is often a colorless fluid, although hemocytes within serum may impart color to it. The fluid is iso-osmotic with seawater, generally neutral in pH and contains low concentrations of sulfate and proteins. Serum proteins are not involved in maintaining the osmotic tension of the serum, but appear to play significant roles in defensive responses (Wright 1981). Serum-mediated defensive reactions include agglutination of erythrocytes and opsonization of nonself material, but serum does not clot in response to wounding (Wright 1981).

At the molecular level, a number of vertebrate hormone-like factors such as thyroxin and oxytocin, as well as vasoconstrictors, have been found in serum (Barnes 1980). It is also known to contain molecules that are toxic to potential pathogens. These antimicrobial molecules include clavanins and styelins from

Styela clava, didemnins A, B and C from *Trididemnum solidum*, picavines from *Clavenia picta*, lissoclinotoxins A and B from *Lisoclimun perforatum*, aplidium 1,2,3-trithiane from *Aplidium sp.*, halocyamine A and B from *H. roretzi* and Factors 1 and 2 from *Ciona intestinalis* (reviewed in Findlay and Smith 1995).

1.4.2 Hemocytes

The hemocytes within hemolymph are mesodermal in origin. Even free-swimming larvae possess a functional blood vascular system with circulating hemocytes, and morphologically distinct hemocyte types are visible in late larval stages (Wright 1981). Electron and light microscopic observations of hemocytes have revealed numerous morphologically distinct hemocyte subpopulations in adults (Wright 1981; Radford et al. 1998; Fig. 2). Despite the variant terminologies used, Wright (1981) suggests that up to ten distinct cell

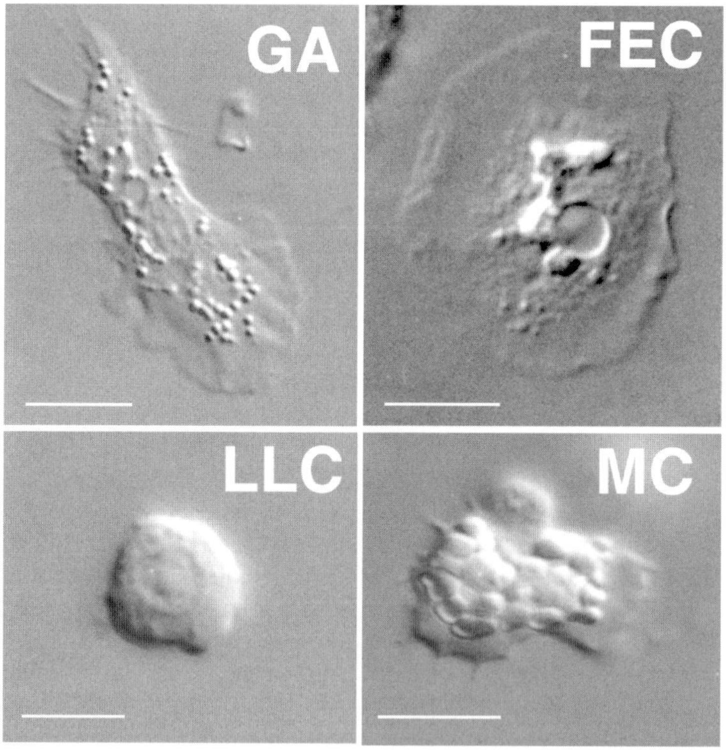

Fig. 2. Phase interference micrographs of hemocytes from the solitary tunicate *Styela plicata*. *GA* granular amebocyte (*bar* 5 µm); *FEC* fried egg cell (*bar* 10 µm); *LLC*, lymphocyte-like cell (*bar* 5 µm); *MC* morula cell (*bar* 10 µm)

types can be identified. These are hemoblasts, lymphocyte-like cells, granular and hyaline amebocytes, vacuolated cells, signet ring cells, compartment cells, morula cells, pigment cells and nephrocytes.

The generalized roles of hemocytes in adults include coagulation, excretion, nutrition, asexual budding, gamete formation, tunic formation and defensive reactions (reviewed in Wright 1981). Hemocytes express a number of defensive lectins, enzymes and antibacterial peptides. These molecules participate in the recognition and destruction of foreign material. Hemocytes are also involved in inflammatory reactions. For instance, the injection of abiotic and biotic material into the tunic induces the massive infiltration of hemocytes. During these inflammation responses, foreign bodies are rapidly sequestered by phagocytosis or encapsulated by hemocytes (Wright 1981). The choice between these different outcomes depends on the size of the foreign material. Small particles injected into the tunic or visceral cavity are efficiently phagocytozed, and the ingested material is then eliminated by intracellular digestion or released into the external environment via the branchial sac, tunic, intestine or neural gland (Wright 1974; Parrinello et al. 1977, 1984; Parrinello and Patricolo 1984). Larger particles are encapsulated and sequestered.

In addition to their phagocytic and encapsulating activities, hemocytes can mount cytotoxic responses against mammalian erythrocytes and nucleated cell lines (Parrinello and Arizza 1992; Parrinello et al. 1993; Peddie and Smith 1993, 1994; Raftos and Hutchinson 1995). Hemocytes from *S. plicata* are also involved in allograft reactions where they appear to undertake cytotoxic reactions towards allogenic tissues (Raftos et al. 1987a,b). Peddie and Smith (1994) suggest that these cytotoxic reactions are preceded by a number of cellular events, including enhanced energy metabolism, cytoskeletal rearrangements and activation of secretory pathways.

1.4.3 The Role of Hemolymph in Regulating Cellular Activity

Even though tunicate serum contains numerous proteins, some with demonstrable defensive functions, humoral regulatory molecules with activities akin to vertebrate cytokines have only been implicated in the regulation of three defense-related cellular responses – inducible cell proliferation, phagocytosis and chemotaxis.

2
Regulation of Cell Proliferation by Humoral Factors

2.1
The Role of Cell Proliferation in Host Defense

A number of studies focusing on histocompatibility responses have suggested that enhanced cell proliferation occurs after immunological challenge in tunicates (Raftos et al. 1987a,b; Raftos and Cooper 1991). Raftos and Cooper (1991) have demonstrated that mitosis in hematopoietic crypts of the body wall in *S. clava* is enhanced by allogenic stimulation. It is likely that this increased proliferation generates new hemocytes, which enter the circulation and contribute to the cytolytic destruction of foreign cells (Raftos et al. 1987b; Kelly et al. 1992a).

2.2
Hematopoiesis in Tunicates

Hematopoiesis in tunicates is well defined. In juvenile *C. intestinalis*, hemocytes in the pharynx divide and migrate, forming the hematopoietic tissues found in the adult pharynx and connective tissues surrounding the gut. These juvenile hemocytes also transform into germinal cells of the gonads (Ermak 1976). After metamorphosis, distinct hematopoietic sites are visible in the connective tissues surrounding the alimentary canal, the transverse bars of the pharynx and the body wall (Wright 1981). The hematopoietic tissues may be organized into distinct nodules, or appear diffused. The locations of these tissues in different tunicate species vary depending on the distribution of other tissues and organs, such as the postpharyngeal gut and gonads (Ermak 1974, 1975, 1976, 1982).

In *S. clava*, distinct hematopoietic nodes are located in the pharynx, body wall and the endocarps, while diffuse patches of hematopoietic tissues are found around the gut (Ermak 1976; Fig. 3). Division of hemoblasts, as measured by the uptake of ^{3}H-thymidine, appears to occur in the hematopoietic tissues before maturation of the hemocytes (Ermak 1974, 1975). However, mitotic division of hemocytes in general circulation has also been observed in *S. plicata* and *S. clava* (Ermak 1975, 1976; George 1939). Renewal of *S. clava* hemocytes generally takes several weeks (within 60 days) (Ermak 1976).

Microscopic observations suggest that all hemocytes are "monophyletic" in origin. Hemoblasts appear to be the general precursor stem cells for all other cell types (Wright 1981). The acquisition of granules and vacuoles differentiates the multipotent hemoblasts into the other cell types. Labeling with ^{3}H-thymidine has provided firm evidence for this scheme of hemocyte differentiation. ^{3}H-thymidine is initially associated with hemoblasts located in the interior of hematopoietic nodules, suggesting that the hemoblasts are actively

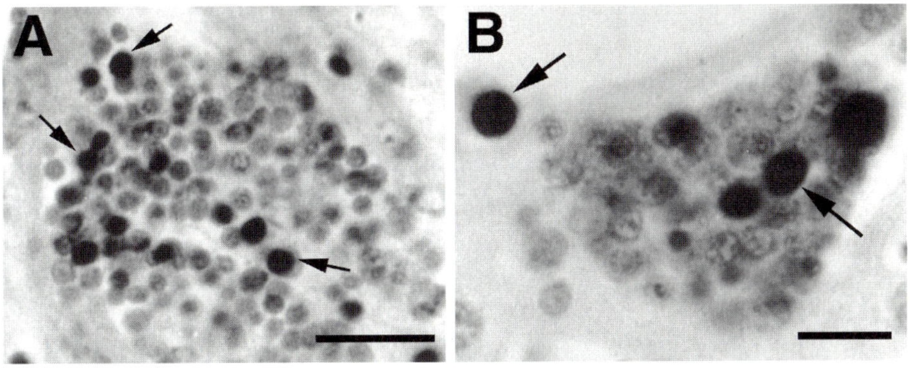

Fig. 3. Bright-field micrographs of hematopoietic crypts in the body wall of *Styela clava*. Body wall tissue was incubated with bromo-deoxyuridine (BrDu), so that proliferating stem cells could be identified by immunostaining with anti-BrDu. Examples of anti-BrDu-stained, proliferating stem cells are identified by *arrows*. **A** *bar* 20 μm; **B** *bar* 10 μm

synthesizing DNA prior to mitotic division (Ermak 1976). After 20 days, labeled cells are located in the periphery of the nodules, from where they enter the general circulation. At this stage, the hemoblasts in the center of the hematopoietic nodules are no longer labeled. This indicates that the hemoblasts undergo differentiation and maturation to form the other hemocyte types located at the periphery of the hematopoietic tissue.

Electron microscopic observations of hematopoietic crypts have shown that such differentiation and maturation involves changes in the nuclei and cytoplasm of the hemocytes (Ermak 1976). For instance, hemoblasts in the center of hematopoietic nodules possess large nuclei with prominent nucleoli. Their cytoplasm contains few distinguishing features, although abundant polyribosomes are visible. Toward the periphery of nodules, the nuclei of the hemocytes appear to be devoid of nucleoli. This may result from an increase in the amount of chromatin within nuclei.

There is also evidence that hemocytes can mature in the general circulation outside the hematopoietic tissues. Peripheral hemocytes acquire electron-dense granules in their cytoplasm, suggesting that they accumulate stored molecules in membrane-bound vesicles (Ermak 1976). Granular amebocytes, signet ring cells, compartment cells, morula cells and pigment cells all seem to differentiate in the hemolymph (Wright 1981).

2.3 Humoral Mitogens

The first direct indication that hematopoiesis was sensitive to humoral regulation came from studies using mammalian cytokines. Raftos et al. (1991) have shown that proliferation in the pharyngeal hematopoietic tissues of *S. clava* can be significantly enhanced by culture with human recombinant IL-2 (hrIL-2). Increased ^{3}H-thymidine uptake was detected among cells in pharyngeal tissue explants that had been incubated with hrIL-2. Responses to hrIL-2 were dose-dependent and seemed to affect lymphocyte-like cells. Enhanced proliferation was stimulated by hrIL-2 in the absence of costimulants (Raftos et al. 1991). Mitogenesis was not evident in pharyngeal explants cultured with other cytokines, such as hrIL-1, that are also known to stimulate proliferation in mammals. This suggests that hrIL-2's effects in tunicates are relatively specific, and implies that tunicate cells have surface receptors that can specifically bind IL-2 and initiate subcellular responses (Raftos et al. 1991).

Such conclusions must, however, be tempered by considerable caution. It is difficult to accept that cytokines like IL-2 have been sufficiently conserved over the course of evolution to allow the mammalian versions of these molecules to bind tunicate cell surface receptors. Even though IL-2-like molecules have been identified in numerous vertebrates, mammalian IL-2 rarely cross-reacts with IL-2 receptors in animals like amphibians and fish (Raftos et al. 1991). Reservations about the molecular basis of IL-2 mediated mitogenesis in tunicates must continue until bona fide homologues of cytokine receptors have been fully characterized in these animals, and until other explanations for IL-2's activities in tunicates have been investigated. It is possible, for instance, that IL-2 activates tunicate cell proliferation without binding to highly specific IL-2 receptors. IL-2 is known to have lectin-like properties (Sherblom et al. 1989), and there are numerous studies demonstrating that the engagement of cell surface carbohydrates by lectins can activate cell proliferation (see below).

Despite these concerns about the effects of mammalian cytokines on tunicate cells, it is clear that tunicates themselves possess humoral proteins with the capacity to regulate mitogenesis. Serum from two species of styelid tunicate, *S. clava* and *S. plicata*, can stimulate the proliferation of cells in pharyngeal explants (Nair et al. 2001; Raftos et al. 1991). In both species, mitogenic activity is strictly dependent on serum concentration, ruling out the possibility that proliferation is stimulated by some unknown component in pharyngeal cultures other than serum.

The mitogenic factors responsible for enhanced proliferation have been purified from the sera of *S. clava* and *S. plicata* using a combination of isoelectric focusing and/or gel filtration (Nair et al. 2001; Raftos et al. 1991). In both species, the mitogens were found to be low molecular weight monomeric proteins. In *S. clava*, the mitogenic protein, designated tunIL-1β, has an approximate molecular weight of 17 kDa (D. Raftos, unpubl. data), whilst the *S. plicata* mitogen is 14 kDa. Both of these proteins are powerful mitogens, stimulating pharyngeal cell proliferation at concentrations as low as 150 ng/ml (Raftos et al. 1991; Fig. 4).

Interestingly, both proteins, and another molecule from *S. clava* designated tunIL-1α, can also stimulate the proliferation of mouse thymocytes and L-929 fibroblasts (Nair et al. 2001; Raftos et al. 1991). Mitogenic activity towards

Fig. 4. The effect of various concentrations of tunIL-1β on cell proliferation (^{3}H-thymidine uptake ± SEM, $n > 3$) in pharyngeal tissue explants from *Styela clava*

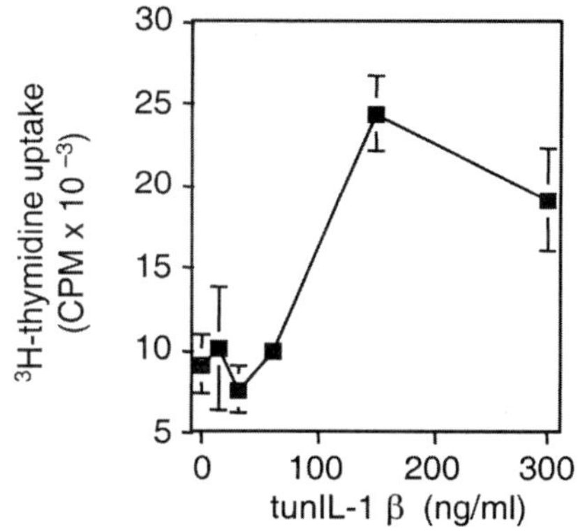

mammalian cells has also been associated with a galactosyl-binding lectin from *Didemnum candidum* (Vasta et al. 1986), and with sera from a number of other tunicates (Beck et al. 1989). The ability of these molecules to modulate mammalian cell proliferation implies that the cellular determinants with which the tunicate proteins interact to initiate subcellular effects have been highly conserved over the course of chordate evolution (a conclusion that evokes the same caution as that for hrIL-2, described above). It has also been demonstrated that the secretion of tunIL-1 proteins from *S. clava* hemocytes can be enhanced by culture with the inflammatory agent, zymosan (Raftos et al. 1992). This capacity for induction indicates that tunIL-1 proteins can adaptively modulate cellular activities in response to infection or inflammation.

3 Humoral Phagocytic Regulators

3.1 Phagocytosis in Tunicates

In addition to their mitogenic activities, tunIL-1 proteins from *S. clava*, and the 14kDa mitogen from *S. plicata*, have been shown to modulate the phagocytic activities of tunicate hemocytes. Phagocytosis has been thoroughly documented in many tunicates, including *C. intestinalis* (Wright and Cooper 1975), *Molgula manhattensis* (Anderson 1971) and *B. schlosseri* (Ballarin et al. 1994). There is some evidence that only a certain proportion of hemocytes is phagocytic (reviewed in Parrinello and Cammarata 1995). Granular, hyaline and vacuolated hemocytes are the main cell types implicated in phagocytosis. While these cells may be considered to be 'professional' phagocytes, other cell types can also ingest nonself particles when the latter are present at high concentrations (high target:hemocyte ratio).

Changes in the morphology of hemocytes from *B. schlosseri* have been studied after the ingestion of nonself material (Cima et al. 1996). Phagocytes change their overall shape from spherical/ovoid to flattened. They also increase the production of superoxide, nitrite ions and acid phosphatase. These biochemical changes probably reflect the activation of specific killing mechanisms (Ballarin et al. 1994; Cima et al. 1996).

Phagocytosis of yeast by *B. schlosseri* hemocytes is significantly depressed in the presence of a variety of pharmacological agents that inhibit Ca^{2+} homeostasis (Ballarin et al. 1997). Phagocyte cell surface receptor engagement appears to activate subcellular signaling pathways resulting in the formation of inositol 1,4,5-triphosphate by phospholipase C. Inositol 1,4,5-triphosphate then mobilizes Ca^{2+} from subcellular stores like the endoplasmic reticulum. The rise in the intracellular concentration of Ca^{2+} initiates reactions such as phagocytosis and superoxide formation (Ballarin et al. 1997).

Phagocytic signaling systems can be activated either by direct interaction of target particles with phagocyte surfaces (non-opsonophagocytosis) or after the opsonization of targets by humoral factors. Non-opsonophagocytosis often results from lectin-carbohydrate interactions, in this case via cell-surface lectins (Arizza et al. 1993; Cammarata et al. 1993). The presence of lectins on hemocyte surfaces has been demonstrated by a number of methods, including inhibition of surface binding by carbohydrates, flow cytometric analysis and direct microscopic visualization using immunofluorescence techniques (Vasta 1991). Interestingly, identical lectins in serum and on hemocyte surfaces have been identified in some tunicates (Vasta et al. 1982). Membrane-bound lectins seem to differ from their soluble counterparts in the possession of a membrane anchor. It seems likely that these anchor domains are proteolytically cleaved from surface lectins that are destined to be released into the serum.

Despite their capacity for non-opsonophagocytosis, tunicates also utilize humoral opsonins to enhance phagocytic potential. Incubation of target particles with *H. roretzi* serum enhances phagocytosis by approximately 100% (Ohtake 1994). This can be attributed to the presence of opsonic factors in the serum. Similarly, divalent cation-dependent carbohydrate-specific opsonization has been demonstrated in *B. schlosseri* (Ballarin et al. 1994) and *S. clava* (Kelly et al. 1992b, 1993a). In these species, incubation of target cells with serum increases the rate of phagocytosis by hemocytes as well as increasing the proportion of phagocytic cells. The opsonin in *S. clava* appears to be species-specific, since incubation of target particles with *S. plicata*, as opposed to *S. clava*, serum resulted in a significantly lower rate of phagocytosis by *S. clava* hemocytes. Most reports suggest that opsonins in tunicate serum are carbohydrate-specific lectins. It is also evident, however, that not all tunicate lectins function as opsonins. Although *S. plicata* serum possesses an agglutinin for rabbit erythrocytes, preincubation with serum does not enhance phagocytosis of erythrocytes (Fuke and Sugai 1972).

3.2 Regulation of Phagocytosis by Serum Proteins

The ability of the low-molecular-weight styelid proteins (tunIL-1s from *S. clava* and the 14kDa *S. plicata* protein) to modulate phagocytosis has been revealed in two ways. Beck et al. (1993) have demonstrated that incubating hemocytes with as little as 300ng/ml of tunIL-1α or β from *S. clava* can enhance phagocytic activity, even when these proteins are washed clear of phagocytes prior to addition of phagocytic target cells. This suggests that the styelid proteins bind phagocyte surface molecules that have the capacity to activate phagocytic cells and enhance the subsequent uptake of foreign targets.

Interestingly though, the styelid humoral molecules can also act as opsonins. Both of the tunIL-1 proteins, and the 14kDa mitogen from *S. plicata*, can bind onto the surface of foreign target cells like yeast, and opsonize those cells for phagocytosis (Kelly et al. 1992b; Beck et al. 1993; Nair et al. 2001). This capacity for opsonization is independent of the ability of the proteins to directly activate phagocytes in the absence of target cells. The clear implication of their capacity for opsonization is that the styelid proteins incorporate binding sites for nonself determinants. Additional experiments have shown that these nonself binding sites are comparable to the carbohydrate recognition domains of lectins. Opsonization can be specifically inhibited by galactosyl carbohydrates like galactose and *N*-acetyl-D-galactosamine (Kelly et al. 1992b; Nair et al. 2001). A broad range of other sugars has no inhibitory effect, indicating that the styelid proteins act as galactosyl-specific lectins that opsonize phagocytic targets by binding galactosyl moieties on foreign cells.

4 Regulation of Hemocyte Motility by Humoral Factors

4.1 Chemotaxis and Inflammation in Tunicates

Chemotaxis is the only other defense-related cellular activity, apart from cell proliferation and phagocytosis, that is known to be influenced by humoral regulatory proteins. A number of studies have demonstrated that tunicate hemocytes home to sites of antigenic challenge, implying a capacity for directed migration (chemotaxis). For instance, large aggregates of hemocytes form around, and migrate into, allogenic tunic transplants in *S. plicata* (Raftos et al. 1987b; Raftos 1996a). Autografts do not exhibit the same level of cellular infiltration, suggesting that the specific recognition of alloantigens stimulates cellular recruitment.

Similar directed migration can also be generated by pathogen-associated antigens. The injection of particulate (erythrocytes, oocytes, stromata or colloidal carbon) or soluble (BSA, hemoglobin or hemocyanin) antigens into the epidermis of tunicates leads to the active infiltration of macrophage-like hemocytes. Infiltrating hemocytes phagocytose or encapsulate the injected material and undergo rapid degranulation. Degranulation presumably releases a variety of defense-related molecules from intracellular stores (Parrinello and Patricolo 1984; Parrinello et al. 1984; Wright 1974). Whilst observations of this type have been available since the early work of Metchnikoff in the 19th century (Metchnikoff 1892), humoral molecules that direct the migration of defensive cells to sites of inflammation have only recently been identified.

4.2 Tunicate Chemotaxins

Again, tunIL-1 proteins from *S. clava* appear to be involved in chemotactic activity by tunicate hemocytes (Raftos et al. 1998). Both *S. clava* serum and purified tunIL-1 proteins can enhance the migration of hemocytes in vitro (Kelly et al. 1993b; Raftos et al. 1998). Checkerboard analyses have demonstrated that this capacity to alter migratory activity reflects truly directional chemotaxis (Raftos et al. 1998). It does not result from the simple enhancement of random, nondirectional movement (usually termed chemokinesis). In other words, *S. clava* hemocytes specifically migrate up concentration gradients of tunIL-1 proteins.

Chemotaxin-mediated responses have also been identified in the solitary tunicate, *Pyura stolonifera* (Raftos et al. 2003). Activation of *P. stolonifera* serum with inflammatory agents, such as lipopolysaccharide or zymosan, generates powerful chemotactic activity among hemocytes. Fractionation of zymosan-activated *P. stolonifera* serum has demonstrated that enhanced hemocyte migration is specifically associated with the proteolytic formation of an 8.5 kDa peptide in serum (Fig. 5).

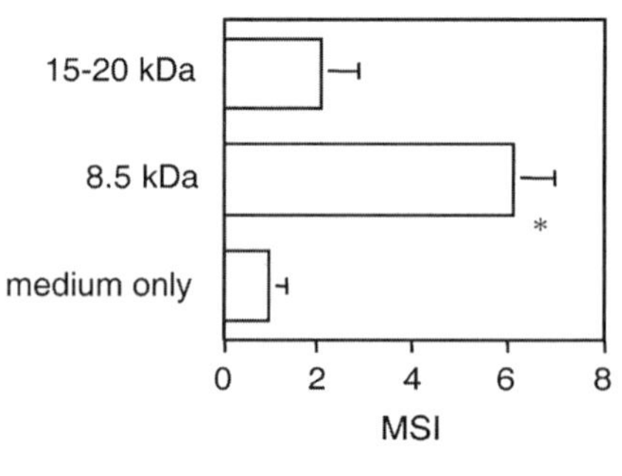

Fig. 5. The effects of an 8.5-kDa anaphylatoxin-like peptide, which was isolated from activated tunicate serum, on the migratory activity of *Pyura stolonifera* hemocytes (migration stimulation index, MSI ± SEM, $n > 5$). Also shown are the effects on migration of tissue culture medium alone and *P. stolonifera* proteins in the size range 15–20 kDa. *, $p < 0.05$ vs. medium alone

5 Molecular Characterization of Humoral Regulatory Factors from Tunicates

5.1 TunIL-1 Proteins from *S. clava*

As their name implies, tunIL-1 proteins from *S. clava* have been favorably compared to the pro-inflammatory vertebrate cytokine, IL-1. Analogies between tunIL-1 and IL-1 have been based on their shared functional characteristics such as mitogenic activity, phagocytic activation and the ability to alter cell migration. They also share a number of physicochemical similarities. Mammalian IL-1 and *S. clava* tunIL-1 have comparable molecular weights (approximately 18 kDa), and both are expressed as two isoforms (α and β) that can be distinguished on the basis of their isoelectric points (Beck et al. 1989; Beck and Habicht 1991). In addition, anti-mammalian IL-1 antibodies seem to cross react with tunIL-1 proteins, as well as with molecules from *B. schlosseri* (Beck et al. 1993; Ballarin et al. 2001). In situ hybridization has also identified mRNAs in *S. plicata* that hybridize with mammalian IL-1β probes (Pestarino et al. 1997).

However, despite their similarities, there is no definitive evidence for evolutionary homology between tunIL-1s and vertebrate IL-1s. All of the existing data can be equally well explained by chance structural or functional similarities between tunIL-1s and IL-1s, or by convergent evolution. *S. clava* tunIL-1 proteins have not been characterized at the molecular level, and so their true evolutionary relationships remain obscure.

5.2 14 kDa Mitogen/Opsonin from *S. plicata*

The same is not true for the 14 kDa mitogen/opsonin isolated from *S. plicata* serum. This protein is of comparable molecular weight to *S. clava* tunIL-1's and it mediates many of the same cellular functions. It also undertakes a number of additional IL-1-like functions. It can increase IL-2 secretion by mammalian peripheral blood mononuclear cells and it enhances IL-2 receptor (IL-2R) expression by mammalian EL-4.IL-2 cells (Nair et al. 2001).

However, molecular characterization of the *S. plicata* protein suggests that it is not an IL-1 homologue. N-terminal and internal amino acid sequencing have deciphered the first 49 amino acids of the *S. plicata* molecule. This represents up to 50% of the entire protein. ClustalW alignments of this sequence with consensus sequences for the two subfamilies of IL-1 (IL-1α and IL-1β/IL-1ra) failed to reveal any significant homology, even though the N-termini of both IL-1 subfamilies contain a number of highly conserved amino acids. The

IL-1α subfamily, for instance, is characterized by numerous absolutely conserved amino acids within its N-terminus (S^1, Y^{12}, N^{26}, D^{27}, L^{29}, N^{30}, Q^{31}, L^{42}) (Nair et al. 2001).

Instead of being homologous to IL-1, the 14kDa *S. plicata* protein has very strong similarities to a number of 14kDa lectins (TC-14s) from a closely related tunicate, *Polyandrocarpa misakiensis* (Suzuki et al. 1990). TC-14s belong to a group of C-type (Ca^{2+}-dependent) lectins with carbohydrate recognition domains that specifically bind galactosyl carbohydrates. Sixty seven percent of amino acids in the 14kDa *S. plicata* sequence are identical, or are conservatively substituted, with those in the *P. misakiensis* TC14 proteins (Nair et al. 2001). This obvious homology to galactose-specific, Ca^{2+}-dependent lectins explains much of the functional data associated with the *S. plicata* protein. The *S. plicata* molecule is known to opsonize target cells for phagocytosis. Opsonization can be inhibited by galactosyl sugars and EDTA, suggesting that its capacity for target recognition results from the Ca^{2+}-dependent binding of galactosyl carbohydrates. Moreover, the TC-14 lectins from *P. misakiensis,* to which the *S. plicata* 14kDa molecule is closely related, are known to act as critical cytophilic regulators. They control the growth, adhesion and differentiation of multipotent epithelial cells in budding tunicates, in much the same way that the *S. plicata* protein promotes cellular proliferation (Matsumuro et al. 2001).

The association of the *S. plicata* protein with C-type lectins also helps to resolve incongruities in functional data. Comparative inhibition experiments using mammalian IL-1 have proven that the ability of the 14kDa *S. plicata* protein to stimulate the proliferation of mammalian cells is not mediated by IL-1 receptors (IL-1R). The 14kDa *S. plicata* protein cannot block the binding of anti-IL-1R antibodies to mammalian IL-1R, and visa versa, even though human IL-1α and β very effectively inhibit such interactions (Raftos 1996b). This indicates that the *S. plicata* protein initiates the proliferation of mammalian cells via some other cell surface determinant.

In light of its obvious relationship to C-type lectins, it seems likely that the *S. plicata* protein gains its effects on mammalian cells by interacting with surface galactosyl moieties rather than IL-1-specific receptors. The ability of such surface carbohydrate-based cellular interactions to stimulate proliferation is thoroughly documented. Many studies of mammalian cells have shown that plant and animal lectins with comparable sugar specificities to the *S. plicata* protein are powerful mitogens (Lis and Sharon 1986; Vasta et al. 1986). The *S. plicata* protein's ability to stimulate the proliferation of tunicate cells might be similarly explained, since other lectins are known to induce mitogenesis in *S. clava* (Raftos et al. 1991).

5.3 Complement Component C3-like Anaphylatoxins

The 8.5-kDa peptide from *P. stolonifera* serum that induces the chemotaxis of hemocytes has also been characterized at the molecular level. In this case, the peptide appears to be derived from a homologue of the central complement component, C3 (Raftos et al. 2002, 2003). Prior to studies of *P. stolonifera*, homologues of C3 had been characterized in two other tunicate species, *H. roretzi* and *S. plicata* (Nonaka and Azumi 1999; Nonaka et al. 1999; Smith et al. 1999; Raftos et al. 2002). In both species, C3 homologues act as powerful opsonins that can be activated by Ca^{2+}-dependent lectins in a manner analogous to vertebrate lectin-mediated activation pathways (Smith et al. 1999).

The entire *H. roretzi* C3-like gene has been sequenced (Nonaka et al. 1999). That sequence reveals obvious similarities to C3 molecules from echinoderms and a range of jawed vertebrates. It also identifies a motif in the predicted α-chain that is equivalent to a critical proteolytic cleavage site in vertebrates that initiates downstream effector activity by forming two peptides – the opsonin, C3b, and the 9 kDa anaphylatoxin, C3a.

Very recent evidence suggests that the 8.5 kDa peptide generated by the activation of *P. stolonifera* serum is a homologue of the C3a anaphylatoxin (Raftos et al. 2002b). Amino acid sequence and amino acid composition analysis of the *P. stolonifera* peptide suggest that it is closely related to the anaphylatoxin-like region of the *H. roretzi* C3a chain. It shares 72% amino acid identity with the *H. roretzi* C3a-like sequence in an 18 amino acid overlap. The 8.5-kDa peptide from *P. stolonifera* can also be detected by an anti-C3 antibody that is known to cross react with tunicate C3 molecules.

6 Conclusions and Perspectives

The evidence presented in this chapter suggests that some host defense responses in tunicates are regulated by cytophilic humoral factors. It has been known for some time that many cellular responses in tunicates are induced by infection and other forms of antigenic challenge. In at least three cases – cellular proliferation, phagocytosis and chemotaxis – these inducible responses are clearly modulated by the action of humoral regulatory proteins. However, this chapter also sounds a cautionary note. Even though functional and physicochemical characteristics have suggested that inducible cellular responses in tunicates may be regulated by the evolutionary precursors of vertebrate cytokines, bona fide homologues of cytokines, such as interleukins, tumor necrosis factors or interferons, have not yet been identified in this group of animals. This does not suggest that cytokine homologues will not be found in tunicates. But it is apparent that the search for such molecules will be hampered by the substantial evolutionary distance between tunicates and their most closely related vertebrate cousins.

References

Anderson RS (1971) Cellular receptor to foreign bodies in the tunicate, *Molgula manhattensis* (DeKay). Biol Bull 141:91–98

Arizza V, Parrinello N, Cammarata M, Picciurro A (1993) Immunocytochemical localization of cellular lectins in *Phallusia mamillata* hemocytes. Anim Biol 2:83–90

Ballarin L, Cima F, Sabbadin A (1994) Phagocytosis in the colonial ascidian, *Botryllus schlosseri*. Dev Comp Immunol 18:467–481

Ballarin L, Cima F, Sabbadin A (1997) Calcium homeostasis and yeast phagocytosis in hemocytes of the colonial ascidian *Botryllus schlosseri*. Comp Biochem Physiol 118A:113–119

Ballarin L, Franchini A, Ottaviani E, Sabbadin A (2001) Morula cells as the major immunomodulatory hemocytes in ascidians: evidence from the colonial species *Botryllus Schlosseri*. Biol Bull 201:59–64

Barnes RD (1980) Invertebrate Zoology. Holt-Saunders, Tokyo

Beck G, Habicht GS (1991) Primitive cytokines: harbingers of vertebrate defense. Immunol Today 12:180–183

Beck G, Vasta GR, Marchalonis JJ, Habicht GS (1989) Characterization of interleukin-1 activity in tunicates. Comp Biochem Physiol 92B:93–98

Beck G, O'Brien RF, Habicht GS, Stillman DL, Cooper EL, Raftos DA (1993) Invertebrate cytokines. III: Interleukin-1-like molecules stimulate phagocytosis by tunicate and echinoderm cells. Cell Immunol 146:284–299

Berrill NJ (1950) The Tunicata. The Ray Society, London

Berrill NJ (1955) The origin of the vertebrates. Clarendon Press, Oxford

Cammarata M, Parrinello N, Arizza V (1993) *In vitro* release of lectins from *Phallusia mamillata* hemocytes after their fractionation on a density gradient. J Exp Zool 266:319–327

Cima F, Ballarin L, Sabbadin A (1996) New data on phagocytes and phagocytosis in the compound ascidian *Botryllus schlosseri*. Ital J Zool 63:357–364

Dakin WJ (1980) Australian Seashores. Angus and Robertson, London

Ermak TH (1974) Cell proliferation in the digestive tract of *Styela clava* as revealed by autoradiography with tritiated thymidine. J Exp Zool 194:449–466

Ermak TH (1975) An autoradiographic demonstration of blood cell renewal in *Styela clava* (Urochordata : Ascidiacea). Experientia 31:837–839

Ermak TH (1976) The hematogenic tissues of tunicates. In: Wright RK, Cooper EL (eds) Phylogeny of thymus and bone marrow-bursa cells. Elsevier/North Holland, Amsterdam, pp 45–56

Ermak TH (1982) The renewing cell populations of ascidians. Am Zool 22:795–805

Field CG, Olsen GJ, Lane DJ, Giovannoni SJ, Ghiselin MT, Raff EC, Pace NR, Raff RA (1988) Molecular phylogeny of the animal kingdom. Science 239:748–753

Findlay C, Smith VJ (1995) Antimicrobial factors in solitary ascidians. Fish Shellfish Immunol 5:645–658

Fuke M, Sugai T (1972) Studies on the naturally occurring hemagglutinin in the coelomic fluid of an ascidian. Biol Bull 143:140–149

Garstang W (1928) The morphology of the Tunicata, and its bearing on the phylogeny of the Chordata. Q J Microsc Sci 72:51–187

George WC (1939) A comparative study of the blood of tunicates. Quart Rev Microscop Sci 81:391–431

Kelly K, Cooper EL, Raftos DA (1992a) In vitro allogeneic cytotoxicity in the solitary urochordate, *Styela clava*. J Exp Zool 262:202–208

Kelly K, Cooper EL, Raftos DA (1992b) Purification and characterisation of a humoral opsonin from the solitary urochordate, *Styela clava.* Comp Biochem Physiol 103B:749–753

Kelly K, Cooper EL, Raftos DA (1993a) A humoral opsonin from the solitary urochordate, *Styela clava.* Dev Comp Immunol 17:29–39

Kelly K, Cooper EL, Raftos DA (1993b) Cytokine-like activities of a humoral opsonin from the solitary tunicate *Styela clava.* Zool Sci 10:57–64

Lis H, Sharon N (1986) Biological properties of lectins. In: Liener IE, Sharon N, Goldstein IJ (eds) The lectins: properties, and applications in biology and medicine. Academic Press, Orlando, pp 266–285

Matsumuro J, Nakamoto C, Fujiwara S, Yubisui T, Kawamura K (2001) A novel C-type lectin regulating cell growth, cell adhesion and cell differentiation of the multipotent epithelium in budding tunicates. Development 128:3339–3347

Medzhitov R, Janeway CA (1997) Innate immunity: the virtues of a nonclonal system of recognition. Cell 91:295–298

Metchnikoff E (1892) Lectures on the comparative pathology of inflammation. Dover, New York

Nair S, Burandt M, Hutchinson A, Raison R, Raftos D (2001) A C-type lectin from the tunicate, *Styela plicata*, that modulates cellular activity. Comp Biochem Physiol 129C:11–24

Nonaka M, Azumi K (1999) Opsonic complement system of the solitary ascidian, *Halocynthia roretzi.* Dev Comp Immunol 23:421–427

Nonaka M, Azumi K, Ji X, Namikawa-Yamada C, Sasaki M, Saiga H, Dodds AW, Sekine H, Homma MK, Matsushita M, Endo Y, Fujita T (1999) Opsonic complement component C3 in the solitary ascidian, *Halocynthia roetzi.* J Immunol 162:387–391

Ohtake S (1994) The phagocytes in haemolymph of *Halocynthia roretzi* and their phagocytic activity. Zool Sci 11:681–691

Parrinello N, Arizza V (1992) Cytotoxic activity of invertebrate hemocytes with preliminary findings on the tunicate *Ciona intestinalis.* Boll Zool 59:183–189

Parrinello N, Cammarata M (1995) Collection, separation, identification and phagocytic activity of tunicate hemocytes. In: Stolen J (ed) Techniques in fish immunology: immunology and pathology of aquatic invertebrates. SOS Publ, Boca Raton, pp 1–15

Parrinello N, Patricolo E (1984) Inflammatory-like reaction in the tunic of *Ciona intestinalis* (Tunicata) II. Capsule formation. Biol Bull 167:238–250

Parrinello N, Patricolo E, Canicatti C (1977) Tunicate immunobiology I. Tunic reaction of *Ciona intestinalis* L. to erythrocyte injection. Boll Zool 44:373–381

Parrinello N, Patricolo E, Canicatti C (1984) Inflammatory reaction in the tunic of *Ciona intestinalis* (Tunicata) I. Encapsulation and tissue injury. Biol Bull 167:229–237

Parrinello N, Arizza V, Cammarata M, Parrinello DM (1993) Cytotoxic activity of *Ciona intestinalis* hemocytes: properties of the *in vitro* reaction against erythrocyte targets. Dev Comp Immunol 17:19–27

Peddie CM, Smith VJ (1993) *In vitro* spontaneous cytotoxic activity against mammalian target cells by hemocytes of the solitary ascidian, *Ciona intestinalis.* J Exp Zool 269:616–623

Peddie CM, Smith VJ (1994) Mechanism of cytotoxic activity by hemocytes of the solitary ascidian, *Ciona intestinalis.* J Exp Zool 270:335–342

Pestarino M, Anna ED, Masini M, Sturia M (1997) Localisation of interleukin-1 beta mRNA in the cerebral ganglion of the protochordate, *Styela plicata.* Neurosci Lett 222:151–154

Radford JL, Hutchinson AE, Burandt M, Raftos DA (1998) A hemocyte classification scheme for the tunicate *Styela plicata* (Lesueur, 1823). Acta Zool 79:343–350

Raftos DA (1996a) Histocompatibility reactions in invertebrates. Adv Comp Environ Physiol 24:77–121

Raftos DA (1996b) Interactions of tunicate immunomodulatory proteins with mammalian cells. Immunol Cell Biol 73:26–31

Raftos DA, Cooper EL (1991) Proliferation of lymphocyte-like cells from the solitary tunicate, *Styela clava,* in response to allogeneic stimuli. J Exp Zool 260:391–400

Raftos DA, Hutchinson A (1995) Cytotoxicity reactions in the solitary tunicate *Styela plicata.* Dev Comp Immunol 19:463–471

Raftos DA, Raison RL (1992) Out of the primordial slime: evolution and the immune system. Today's Life Sci 14:16–20

Raftos DA Tait NN, Briscoe DA (1987a) Allograft rejection and alloimmune memory in the solitary urochordate, *Styela plicata.* Dev Comp Immunol 11:343–351

Raftos DA, Tait NN, Briscoe DA (1987b) Cellular basis of allograft rejection in the solitary urochordate, *Styela plicata*. Dev Comp Immunol 11:713–726

Raftos DA, Cooper EL, Stillman DL, Habicht GS, Beck G (1991a) Tunicate cytokines: tunicate cell proliferation stimulated by an endogenous hemolymph factor. Proc Natl Acad Sci USA 88: 9518–9522

Raftos DA, Stillman DL, Cooper EL (1991b) Interleukin 2 and phytohemagglutinin stimulate the proliferation of tunicate cells. Immunol Cell Biol 69:225–234

Raftos DA, Cooper EL, Stillman DL, Habicht GS, Beck G (1992) Invertebrate cytokines: release of IL-1-like molecules from tunicate hemocytes stimulated with zymosan. Lymphokine Cytokine Res 11:235–240

Raftos DA, Stillman DL, Cooper EL (1998) Chemotactic responses of tunicate (Urochordata: Ascidiacea) hemocytes *in vitro*. J Invert Pathol 72:44–49

Raftos D, Nair S, Robbins J, Newton R, Peters R (2002) A complement component C3-like protein from the tunicate, *Styela plicata*. Dev Comp Immunol 26:307–312

Raftos D, Robbins J, Newton R, Nair S (2003) A complement component C3-like peptide stimulates chemotaxis by hemocyte from an invertebrate chordate – the tunicate *Pyura stolonifera*. Comp Biochem Physiol A 134(2):377-386

Sherblom AP, Sathamoorthy N, Decker JM, Muchmore AV (1989) IL-2, a lectin with specificity for high mannose glycopeptides. J Immunol 143:939–944

Smith LC, Azumi K, Nonaka M (1999) Complement systems in invertebrates. The ancient alternative and lectin pathways. Immunopharmacology 42:107–120

Suzuki T, Takagi T, Furokuohri T, Kawamura K, Nakauchi M (1990) A calcium-dependant, galactose-binding lectin from the tunicate, *Polyandrocarpa misakiensis*. J Biol Chem 265:1274–1281

Vasta GR (1991) The multiple biological roles of invertebrate lectins: their participation in nonself recognition mechanisms. In: Warr GW, Cohen N (eds) Phylogenesis of immune functions. CRC Press, Boca Raton, pp 74–96

Vasta GR, Sullivan TJ, Cheng TC, Marchalonis JJ, Warr GW (1982) A cell membrane associated lectin of the oyster hemocyte. J Invert Pathol 40:367–373

Vasta GR, Marchalonis JJ, Decker JM (1986) Binding and mitogenic properties of a galactosyl-specific lectin from the tunicate *Didemnum candidum* for murine thymocytes and splenocytes. J Immunol 137:3216–3233

Wright RK (1974) Protochordate immunity I. Primary response of the tunicate *Ciona intestinalis* to vertebrate erythrocytes. J Invert Pathol 24:29–36

Wright RK (1981) Urochordates. In: Ratcliffe NA, Rowley AF (eds) Invertebrate blood cells. Academic Press, London, pp 565–626

Wright RK, Cooper EL (1975) Immunological maturation in the tunicate *Ciona intestinalis.* Am Zool 15:21–27

Subject Index

Printing (Computer to Film): Saladruck Berlin
Binding: Stürtz AG, Würzburg